T0200760

PROBLEM SOLVING FOR PROCESS OPERATORS AND SPECIALISTS

PROBLEM SOLVING FOR PROCESS OPERATORS AND SPECIALISTS

Joseph M. Bonem

AIChE®

WILEY

A JOHN WILEY & SONS, INC., PUBLICATION

Published by John Wiley & Sons, Inc., Hoboken, New Jersey
Published simultaneously in Canada

For general information on our other products and services or for technical support, please contact our Customer Care Department within the United States at 877-762-2974, outside the United States at 317-572-3993 or fax 317-572-4002.

Wiley also publishes its books in a variety of electronic formats. Some content that appears in print may not be available in electronic formats. For more information about Wiley products, visit our web site at www.wiley.com.

Library of Congress Cataloging-in-Publication Data:

Bonem, Joseph M.
    Problem solving for process operators and specialists / Joseph M Bonem.
        p. cm.
    Includes index.
    ISBN 978-0-470-62774-7 (cloth)
    1. Chemical engineering–Problems, exercises, etc.   2. Chemical engineering–Quality control.   3. Chemical processes–Mathematical models.   4. Problem solving.   5. Engineering mathematics–Formulae.   I. Title.
    TP168.B66 2011
    660'.28–dc22

                                                                                2010028354

Printed in Singapore

oBook ISBN: 978-0-470-93396-1
ePDF ISBN: 978-0-470-93395-4

10 9 8 7 6 5 4 3 2 1

This book is an attempt both to leave a legacy and to provide assistance to a group that I consider underpaid and undervalued: the process operations department in refineries and chemical plants. This group was invaluable to me when I began a career with a degree in chemical engineering. They taught me things that were not in books. I am hoping that through this book they can develop into true engineering problem solvers.

I am dedicating this book to these men and women who were so instrumental to helping me get started in a career that I found to be more than rewarding. They are all deceased now and their names are too numerous to mention. But I remember their names, faces, and the things that they taught me.

CONTENTS

CONTENTS

PREFACE

During my career, I spent many hours solving chronic and/or severe chemical plant operating problems. These were problems that others, in the judgment of management, did not have the capability to solve. Some of the problems had resisted multiple attempts by others to find a solution. These problems appeared to be difficult to solve for one of two reasons. Those trying to solve them either did not have the capability or did not use the capability that they had to solve the problems. As I examined the reasons that I was successful in problem solving and others were not, I concluded that there was a common thread that linked all the unsuccessful problem solvers. The three aspects of the common thread were as follows:

- There was no disciplined problem-solving approach used by either the operators or the engineers. Thus both professions tended to "jump to conclusions." That is, they selected the first solution to a problem that they heard or thought about. Even worse than that, sometimes the problem that they were solving was really not the problem that was occurring. They did not take time to understand what problem they were trying to solve.
- They either did not have adequate training or did not use the training that they had to formulate theoretically correct working hypotheses. On one occasion, I observed an engineer correlating fractionation operation with the phase of the moon. This was a truly creative idea, but it had no relationship to reality.
- If a solution *appeared* to work, there was a tendency to immediately proclaim that the problem was solved. Many times the problem went away of its own accord only to resurface at a later date. I often heard the expression "We did a couple of things and the problem went away." As I heard this, I normally said either to myself or aloud, "If you don't know why it went away, there is a high probability that it will come back."

During years of experience, I observed well-trained chemical engineers who had graduated in the upper part of their class having difficulty solving technical plant problems. I concluded that often these well-trained engineers were not really trained in subjects that would allow them to solve real-life plant process problems. They had minimal training in problem-solving techniques and much of their academic training was not directed at pragmatic solutions. For example, the academic world was teaching thermodynamics, but not how the theory applies to reciprocating or centrifugal compressors. With this limited amount of training in approaching real-world problems, the process engineer would often settle for a problem-solving approach based on logic with no calculations or, even worse, simple intuition. In addition, the pressures of a real-life problem-solving environment often caused him or her to take the approach of "trying something even if it doesn't work." Management rarely indicated that the engineer should take the time to make sure that the problem was solved correctly. Many times the belief of an operator and/or mechanic was taken as being the correct problem solution simply because the engineer did not have a good framework to develop any other possibility. Since the graduate engineer and operator now seemed to agree, management felt comfortable in implementing the "joint recommendation."

At the other extreme, I often saw well-trained, experienced operators taking one of two approaches when confronted with a problem.

- They would propose a solution that was based on experience only. If the problem was outside their realm of experience, they would withdraw and proclaim that their job was to "turn valves."
- They would propose an elaborate theory that had no scientific basis and was not based on any calculations.

Often, the solutions developed by the plant engineer or operator did not solve the problem. Even worse, the results of the attempted solutions were not documented and there was a strong possibility that the same problem solution would be tried again at a later date.

In an effort to mitigate the failure to solve industrial problems, a new series of techniques were developed that called for using teams to solve problems via interactive "brainstorming" approaches. I observed that the advantages of these teams were that they often brought a tremendous amount of data to bear on the problem and that they generated a long list of possible hypotheses. However, this approach was no more effective than the previous ones. The reason why this large amount of data failed to produce effective solutions was that there was no systematic analysis of the data. In addition, there was no stipulation that the possible hypotheses had to be theoretically feasible. Thus theoretically impossible hypotheses were treated with the same validity as the theoretically possible ones. The most likely outcome of such brainstorming sessions was that the solution with either the most votes or the loudest proponents was adopted as the recommended approach.

In spite of these less than perfect approaches, industrial problems are being solved by intuitive, logical approaches and/or brainstorming that do not involve calculations and/or data analysis. Most of these problems are being solved by experienced engineers and/or operators. However, these problems are generally not complex or chronic in nature. It is the chronic problems and/or those requiring an advanced analysis that this book addresses.

In addition to the factors discussed above, there are four macro demographic and economic trends that are emerging:

- As the "baby boomers" age and retire, the experience that is often of value in problem solving is not being replaced. Thus a more structured approach will become even more important in the next decade.
- In an effort to become more efficient, technical staffing in process plants is being reduced. Thus there is more emphasis on operators and process specialists being able to solve not only typical operating problems, but the severe and chronic problems.
- Process plants themselves are becoming larger and more complicated. Thus problems are more difficult to solve and there is more incentive for expediting a solution.
- Operators are much better schooled in the basic tools of the process industry such as chemistry and mathematics. With this basic training, they can be trained to do many of the more advanced chemical engineering calculations.

These macro trends lead to the premise of this book as follows:

There is a need to provide problem-solving training to a new generation of process operators who can be trained to do basic chemical engineering calculations rather than only relying on experience-based solutions.

I have been encouraged to meet this goal as I have reflected on operators that have attended my problem-solving courses. While these courses were primarily developed for engineers, the operators in attendance could easily follow and participate as the problem-solving concepts were discussed. However, they lagged somewhat when actual calculations were presented. Thus I have included a separate chapter in this book to provide some basic chemical engineering concepts for someone with the equivalent of high school chemistry and algebra training. With this addition to a book or training seminar, it is my belief that a process operator or specialist will be able to comprehend the concepts presented throughout this book.

I cannot claim that the techniques discussed in this book will allow the process operator or specialist to achieve perfection in the area of problem solving. However, I can say that these techniques have worked for me throughout a long career of industrial problem solving. The chemical engineering fundamentals discussed in the book are presented from the perspective of the problem solver as opposed to the perspective of a process designer or that of someone in the academic world. There are shortcuts and simplifying assumptions that are used. These may not be theoretically precise, but they are more than adequate for problem-solving activities. There are without a doubt additional chemical engineering fundamentals that should be covered. However, I have selected those areas that I felt would be of most value to the industrial problem solver.

My industrial experience indicates that there are three requirements to successfully solving complex problems. They are as follows:

1. You must have verifiable data.
2. You must use a structured problem-solving approach that includes a statement of what problem you are trying to solve. This requires rigid discipline. As discussed in Chapter 1, we often fail at simple problem solving because we tend to rely on intuition or experience-based solutions as opposed to a more rigorous, structured problem-solving approach.
3. You must use sound engineering skills to develop a simple working hypothesis.

If any one of these three is absent, unsatisfactory results may occur. For example, a logical solution to a problem is of no value if it is not based on sound data or if the conclusion violates a fundamental premise of engineering. Conversely, a sophisticated computer simulation program is of little value unless it is directed toward solving the correct problem in an expeditious fashion.

Multiple surveys and interviews throughout the United States have listed "problem-solving skills" and "vocational-technical skills" in the top 10 skills that employers wish their employees had. This book deals with these two skills as follows:

1. The three essential problem-solving skills (Daily Monitoring System, Disciplined Problem-Solving Approach, and Determining Optimum Technical Depth) are discussed and guidelines are provided for successful implementation of each of these.
2. Vocational-technical skills are enhanced by equipment descriptions, helpful hints, and practical knowledge that will expand the problem solver's capabilities and academic training. The helpful hints and practical knowledge include calculation techniques that are presented without lengthy derivations and proofs.

Several example problems are included throughout this book in order to illustrate the concepts and techniques discussed. Some of the example problems are included in the chapters devoted to specific aspects of process engineering. The remainder are included in Chapter 14. This chapter is meant to deal with a series of problems that involve multiple aspects of process engineering problem solving.

The problems in the book are, for the most part, real problems. The failures and successes described have actually occurred. The problem-solving techniques described in this book were responsible for the successes. Failures were often due to not using the techniques described. Occasionally, fictitious problems are created to help illustrate important concepts or calculation techniques.

The English set of units has been used throughout the book. The English units and their abbreviations are described at the end of each chapter. A table of conversion factors to scientific units is provided in the Appendix.

Throughout the book, I have used the term "problem solver" to mean the individual with direct responsibility for solving the problem under consideration. I have also used the masculine pronoun "he" knowing full well that there are talented female problem solvers as well.

I have borrowed heavily from my previous book, *Process Engineering Problem Solving*. After publication of the earlier book, I was often asked about the utilization of the principles described in the book by other engineering disciplines or by operators/mechanics. It is my firm belief that the problem-solving principles (Daily Monitoring System, Disciplined Problem-Solving Approach, and Determining Optimum Technical Depth) described in the earlier book can be used by other engineering disciplines or operators/mechanics. This became the incentive to write a book specifically for process operators. An operator cannot be expected to have the academic skills to formulate a full range of process hypotheses. However, because of the pragmatic approach used in this book, it is likely that an operator/mechanic could readily learn how to do the calculations required to formulate a majority of the theoretically correct hypotheses required to solve problems in a process plant. Problem solving throughout this book is referred to as "engineering problem solving" even though the book is intended for operators. The term "engineering problem solving" is used since the proposed approach uses engineering calculations that can be learned by operators.

While this book and my previous book were written for those working in industry, I have been told by others that the early chapters of the books have wide application. These people have suggested to me that these chapters have application to any vocation requiring a disciplined approach to problem solving, such as criminal or fire investigation and medicine.

JOSEPH M. BONEM

INITIAL CONSIDERATIONS

1.1 INTRODUCTION

Problem solving is found throughout all activities of daily life. Problem solving tends to take place in two mind modes. There is the intuitive or instinctive reactionary mode, which has also been called "gut feel." Then there is the methodical reasoning approach, which is usually based on theoretical considerations and calculations.

Both of these approaches have a place in real world problem-solving activities. The intuitive reactionary person will respond much faster to a problem. The response is usually based on experience. That is, he has seen the same thing before or something very similar and remembers what the problem solution was. However, if what is occurring is a new problem or is somewhat different, his approach may lead to an incorrect problem solution. The methodical reasoning person will not be able to react to problems quickly, but will usually obtain the correct problem solution for complicated problems much faster than the intuitive reactionary person, who must develop and perhaps discard several "gut feel" solutions.

Here is an example of how two people with these different mind-sets can react. On a golf course, the cry of "Fore" will elicit different responses. The person responding based on intuition or instinct will immediately crouch and cover his head. This will reduce the probably that the errant golf ball hits a sensitive body part. The person responding based on methodical reasoning will begin to assess where the cry came from and where the ball might be coming

Problem Solving for Process Operators and Specialists, First Edition. Joseph M. Bonem.
© 2011 John Wiley & Sons, Inc. Published 2011 by John Wiley & Sons Inc.

from, and then reach a conclusion as to where it might land. Obviously, in this case, reacting based on intuition or instinct is a far superior mode of operating. There are many more examples from the sports world where reacting in an intuitive fashion yields far superior results than reacting in a methodical reasoning manner. However, essentially all of these examples will be experience-based. People who are reacting successfully in an intuitive mode know what to do because they have experienced the same or very similar situations.

Similar things happen in industrial problem solving. Experienced people such as engineers or operators react instinctively because they have experienced similar events. These operators or engineers do an excellent job of handling emergency situations or making decisions during a startup. As a rule, the person who tends to respond based on methodical reasoning and calculations can rarely react fast enough to be of assistance in an emergency or if quick action is required in a startup situation. The exception to this rule is the engineer who has designed the plant and has gone through calculations to understand what will happen in an emergency or startup. In effect, he has gained the experience through calculations as opposed to actual experience.

The experience necessary to conduct problem solving in the real world does not always exist. In addition, while the need for quick response when solving industrial problems is real, there is not always an emergency or crisis that requires immediate action. Thus the methodical reasoning approach is often the desirable mode of operating. The three components of this methodical reasoning approach are:

1. A systematic, step-by-step procedure. This includes the three essential problem-solving skills (Daily Monitoring System, Disciplined Problem-Solving Approach, and Determining Optimum Technical Depth).
2. A good understanding of how the equipment involved works.
3. A good understanding of the specific technology involved.

Before discussing problem solving in industrial facilities, two examples from everyday life are discussed. It often aids learning to discuss things that are outside the scope of the original thrust of the teaching. The two examples from everyday life discussed below will be helpful in understanding the difference between intuitive problem solving and that based on methodical reasoning.

1.2   AN ELECTRICAL PROBLEM

While trimming bushes with an electric hedge trimmer, a laborer accidentally cut the extension cord being used to power the trimmer. He had been using an electrical outlet in a pump house located approximately 70 ft from the main house. The only other use for 110 volt electricity in the pump house was for a small clock associated with the water softener. The laborer found another extension cord and replaced the severed cord. However, when he plugged it in and tried to turn on the hedge trimmer, it did not have any power. He then had to report the incident to the homeowner. The homeowner checked the panel-mounted circuit breakers. None of them appeared to be tripped. To be sure, he turned off the appropriate circuit breaker and reset it. However, power was still not restored to the outlet in the pump house. To make sure that the replacement extension cord was not the problem, the homeowner plugged another appliance into the electrical outlet in the pump house. It did not work either. The homeowner then concluded that the electric outlet had been "blown out" when the cord was cut. He replaced the electric outlet. However, this still did not provide power to the equipment. When the homeowner rechecked the circuit breaker, he noticed that a ground fault interrupter (GFI) in a bathroom in the main house was tripped. Resetting this GFI solved the problem. Ground fault interrupters are designed to protect from electrical shock by interrupting a household circuit when there is a difference in the currents in the "hot" and neutral wires. Such a difference indicates that an abnormal diversion of current from the "hot" wire is occurring. Such a current might be flowing in the ground wire, such as a leakage current from a motor or from capacitors. More importantly, that current diversion may be occurring because a person has come into contact with the "hot" wire and is being shocked.

While the homeowner believed that in this particular house every GFI protected a single outlet, it is not unheard-of to protect more than a single outlet with a GFI. It seemed surprising that the GFI in a bathroom also protected an outlet in the pump house 70 ft away. The homeowner then recalled that at some point in the past, he had noticed that the small clock in the pump house was about 2 hours slow. This clock was always very reliable. In retrospect, he remembered that at about the same time that the clock lost 2 hours, this particular GFI in the bathroom had tripped during a lightning storm and had not been reset for a few hours. Thus it became obvious that the accidental cutting of the extension cord had caused the GFI to trip rather than tripping the circuit breaker or "blowing out" the electrical outlet. The failure to correctly identify the problem cost the homeowner a small amount of money for the electrical plug and a significant amount of time to go to town to purchase the plug and then install it.

Note that the homeowner's intuitive conclusions were all valid possibilities. That is, the circuit breaker could well have tripped, the replacement extension cord could have had an electrical break in it, or the electrical outlet could have failed when the original extension cord was cut. His problem solving just did not go into enough detail to solve the problem quickly. Several lessons can be learned from this example. While it seemed to be a simple problem that could be easily solved based on the homeowner's experience, the intuitive approach did not work. A more systematic approach based on methodical reasoning might have improved results as follows:

- Consideration would have been given to the possibility that GFIs can protect more than one electrical outlet. The distance between the GFI and the electrical outlet would not be a consideration. The homeowner did not fully understand the technology.

- A voltmeter would have been used to check that power was available coming to the electrical outlet. If power was not available coming to the outlet, the "blown plug" hypothesis would be invalid. A systematic approach was not used.
- In addition, a systematic approach would have raised the question of whether the clock losing 2 hours could be related to the lack of power at the electrical plug.

1.3 A COFFEEMAKER PROBLEM

A man experienced problems with a coffeemaker when it overflowed about half of the time when he made either a flavored or decaffeinated coffee. The coffee and coffee grounds would overflow the top of the basket container and spill all over the counter. The coffee maker performed flawlessly when regular coffee was used. A sketch of the coffee maker is shown in Figure 1-1. When the coffeemaker is started, water is heated and the resulting steam provides a lifting mechanism to carry the mixture of water, steam, and entrained air into the basket where the ground coffee is located. The hot water flows through the coffee and into the carafe. The coffeemaker is fitted with a cutoff valve that causes the flow out of the basket to stop if anyone pulled the carafe out while coffee is still being made.

The man, a graduate engineer, attempted to determine what was wrong. He examined the problem by first convincing himself that he was following

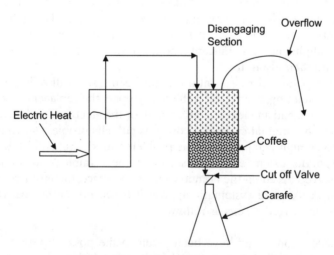

Figure 1-1 Coffeemaker schematic.

directions when it came to making the coffee. He then carefully examined the equipment, especially the cutoff valve. He concluded that somehow the cutoff valve was restricting the liquid flow whenever decaffeinated or flavored coffee was being made. That is, the incoming flow of hot water and steam was greater than the flow out of the valve. This would cause the level in the container to build up and run over. The problem solution seemed relatively simple. He removed the valve and made a sign that read, "Do not remove carafe until coffee is finished brewing." He felt a surge of pride in not only solving the problem, but that he prevented a future problem by providing instructions to prevent someone from pulling out the carafe. The next time that one of the suspect coffees was made, the container did not overflow. He then announced that the problem was solved.

Unfortunately, the glow of successful problem solving did not last long. The next time that flavored coffee was made, the problem recurred; that is, the coffee and grounds flowed over the top of the basket container. The engineer then began a more detailed investigation of the problem, including under-standing the technology for making flavored and decaffeinated coffee. He discovered that when decaffeinated coffee was produced at the coffee supplier, a surface active material was utilized. This surface active material was mixed with the coffee to extract the caffeine. Materials that are surface active have the capability to thoroughly contact the coffee solid so that caffeine is removed from not only the surface, but the deep pores. The surface active material also reduces the surface tension of water, which creates a system that can easily foam.

The engineer then extrapolated from this knowledge and theorized that when flavored coffee was made, a surface active material was used to evenly distribute the flavor to the coffee. Once that he understood the difference in the coffee making processes, he theorized that residual amounts of the surface active material being left on the coffee reduced the surface tension of the hot water and coffee and caused it to foam up in the container and out over the sides onto the counter.

Since the amount of residual surface active material would vary slightly from batch to batch, it was theorized that only the batches of either flavored or decaffeinated coffee that contained greater than a critical level would cause an overflow. After studying this theory, the engineer decided that the problem solution would be to obtain a coffeemaker that had a basket con-tainer with a different design. The problematic coffeemaker had a small cylin-drical basket. A new coffeemaker with a large conical design basket was purchased. The comparison of the two baskets is shown in Figure 1-2. It was theorized that the large conical design would provide a reduced upward velocity of the foaming material and this would allow release of the vapor trapped in the foam. The purchase of this coffeemaker eliminated the problem completely.

Several lessons can be learned from this problem-solving exercise. The intuitive hunch that coffee was not flowing through the valve as fast as hot

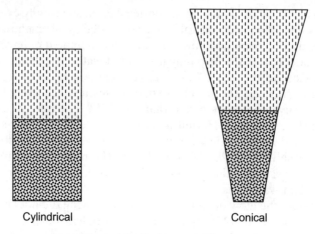

Cylindrical   Conical

Figure 1-2   Basket comparisons.

water was coming into the basket made logical sense. However, no logical explanation was provided for why this only happened with flavored or decaffeinated coffee. Any theory that includes the phrase "for some reason" is suspect and is an indication of an incomplete problem analysis. A portion of an incomplete problem analysis is almost always logical. However, it is imperative that the entire analysis be logical. Another error was that in formulating the hypothesis, the engineer assumed that only liquid water and solid coffee existed in the container. He overlooked the fact that steam vapors and entrained air were always carried into the container with the hot water. The presence of steam and air would provide a mechanism for creating a frothy mixture. The example also illustrates the need for the following:

- A systematic approach, as will be described later in this book, would have eliminated the incomplete hypothesis that suggested the outlet valve was a restriction on only certain grades of coffee.
- A sound understanding of how the equipment works: If the engineer had understood how the coffee maker worked, he would not have assumed that only a liquid was present along with the coffee in the container. He would have recognized that both steam and air were carried over into the container along with the hot water.
- A sound understanding of the technology involved: The fact that decaffeinated and flavored coffee performed differently than did regular coffee should have been an indication to the engineer that he needed to examine the difference in the coffee-making technology.

These relatively simple examples of how successful problem solving requires a more detailed analysis than simple logic and/or intuition are meant to set the stage for the next chapter, which deals with limitations to industrial problem solving. While industrial problems are almost always more complicated than those described in this section, they require the same problem-solving approaches.

1.4 CLASSIFICATION OF INDUSTRIAL PROBLEMS

It will be of value to classify problems into four categories. This will help determine what kind of effort is required to solve the problem. These categories are as follows:

1. *Problems that can be solved based strictly on experience and/or instinct.* These are the problems that are typically solved during a startup and/or upset condition. In these situations, there is minimal time for analysis. Experience and instinct are the only way to solve problems in the time available. As you would imagine, the best problem solvers in this situation are those with experience with the particular problem being encountered.
2. *Equipment problems that can be solved by application of "first principles."* The definition of "first principles" is knowledge that has been summarized as a series of mathematical relationships or expressions. An example of this might be a compressor that is not performing as desired. A study of the head curve (a relationship between flow and pressure head expressed in feet) might reveal that the gas flow and/or molecular weight of the gas have changed so that the anticipated pressures can no longer be achieved.
3. *Process technology problems that can be solved by application of "first principles."* These are process technology problems that can be solved because there are known relationships available. These relationships are often provided in licensing packages or operating instructions. For example, a reactor productivity problem related to impurities in the feed could be solved by using a simplified productivity model. The "base line" for this productivity model will be based on a licensing package or experimental results. The deviation from the base line would be based on laboratory results and a dynamic model. The simple dynamic model would provide a relationship between time and reactor productivity. These models based on the process technology could be used to show that the loss of productivity correlates with "spikes" in a feed contaminant.
4. *Process technology problems that cannot be solved by "first principles."* These might be problems which do not have any reliable solution theory or for which no reliable theory can be developed that relates to the cause of the problem. In other words, theoretically correct "first principles" do not exist. These are usually very complicated problems involving highly

qualitative and subjective variables such as reactor fouling or product attributes such as color, haze, turbidity, or roughness. These subjective variables are often controlled by several independent variables, some of which are well hidden. Some of these controlling variables may be present below the level of analytical detectability. The analysis of such problems is beyond the scope of this book. Fortunately, this classification amounts to only a very small percentage of industrial problems.

The majority of problems in the process industry fall into category 2 or 3. These are the types of problems that the techniques discussed in this book were developed to solve. With an experienced workforce, some of the problems in category 2 or 3 can be solved based on previous history or intuition. However, this experienced workforce is rapidly becoming history. In western countries, the experience level is decreasing as the "baby boomers" reach retirement age. In developing countries, the workforce is just beginning to build experience. Thus there will be an increasing emphasis on using quantitative methods of problem solving. In addition, cost pressures are driving organizations to reduce the number of graduate engineers in operating plants and to use process operators, specialists, or mechanics as the primary problem solvers.

The problem solver will find it helpful to consider which of the above categories best describes a new problem he is trying to solve. This will aid him in determining what kind of effort is required to solve the problem.

2

LIMITATIONS TO PLANT PROBLEM SOLVING

2.1 INTRODUCTION

While later chapters will consider the structured approach to problem solving, any book dealing with plant problem solving will touch on the question, "Is problem solving really part of my job description?" The paradigm of this book is twofold.

- It is based on the concept that all people working in industrial plants have problem solving as part of their job description whether it is written or not.
- To a great extent, the modern process industry has placed operators and process specialists into roles of solving problems. For this problem solving to be done efficiently, they must use some engineering knowledge and calculations. Thus this book discusses "engineering problem solving," meaning problem solving that can be done by engineers or operators using engineering calculations.

The first step in developing an effective problem-solving approach is to have the correct mind-set. Some operators and specialists believe that their job is only to turn valves or make "educated guesses." At the other end of the spectrum some engineers raise the question, "Is problem solving really engineering?" Often, engineers may conclude that problem solving is not truly engineering because of the following:

Problem Solving for Process Operators and Specialists, First Edition. Joseph M. Bonem.
© 2011 John Wiley & Sons, Inc. Published 2011 by John Wiley & Sons Inc.

- Engineering is defined in such narrow terms that only "design work" appears to be engineering.
- Intuition and "gut feel" have replaced thorough analysis as a preferred tool for problem solving.
- Considerations of "optimum technical depth" are not well understood.

If one defines engineering, as dictionaries do, as "The science of making practical application of knowledge in any field," we must conclude that problem solving is truly engineering. In addition, this definition of engineering also fits an operator with engineering training who is working not just to turn valves, but to solve problems.

It is also important to understand why a course in engineering problem solving is of value. In an example of a typical industrial problem, a customer is unhappy with the appearance of the plastic pellets being received from his supplier. Specifically, the pellets have visual discontinuities similar in appearance to gas bubbles. The customer describes these as "voids." If a particle has more than a single void, it is described as a "multi-void particle." A simplified statement of the problem is shown in Figure 2-1. As shown in the figure, the process in which the pellets are manufactured consists of two parts, polymerization and extrusion. In the polymerization section, propylene is polymerized to polypropylene particles (700 microns in diameter) using a catalyst. In the extrusion area, these particles are melted, extruded, and formed into cylinders approximately 1/16 by 1/8 in. A strong correlation was developed between the pellet appearance (fraction of pellets with multi-voids)

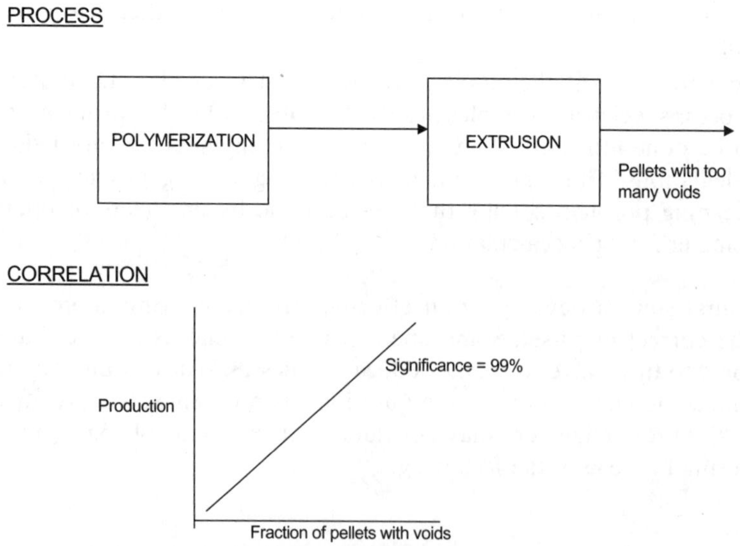

PROCESS

CORRELATION

Figure 2-1   An example of improper problem solving.

and the polymerization production rate. The problem solver recommended that the production rate be reduced to solve the "multi-void" problem. This solution to the problem (reducing production rate) is, at best, only a short-range solution. This solution cannot be considered a lasting solution because of the following:

- The basic cause of the voids was not considered.
- The solution required a severe economic penalty (it might have solved one problem, but it created another one). In most process industries, the limited profits are made at production rates above 75 or 80% of capacity.
- Since the basic cause of the voids was not discovered, the problem will likely recur even at the reduced production rates.

2.2 LIMITATIONS TO PROBLEM SOLVING

The previous example is typical of much of the improper problem solving that occurs in many industries in today's hectic, fast-paced society. It also illustrates why a course in engineering problem solving is of value. There are ten primary limitations to problem solving in today's process plants. They are described as follows.

1. Modern-day processing plants are large and complex. For example, a relatively simple process such as propylene purification has evolved from fractionation followed by a drying process to remove water to a process incorporating "heat pump fractionation" and more complicated conversion steps to remove impurities to the part per billion (ppb) level. In addition, plant sizes have increased significantly. Thus, there is even more emphasis on solving problems quickly and correctly.

2. The problem is usually more complicated than first described. Typical initial problem descriptions might consist of such statements as, "It won't work as designed," or, "It won't work unless you modify it to …" If either of these problem descriptions is accepted exactly as stated, the problem solver is doomed to failure. In order to practice true engineering problem solving, the problem solver must use a disciplined approach that involves writing out an accurate description of the problem that does not include a problem solution. This is necessary to avoid ignoring data and jumping to conclusions.

3. Conflicting data will always be present, and can take many forms. Some examples are that the verbal descriptions eyewitnesses give can disagree; laboratory data may be in disagreement with physical factors, instrumentation, or even other laboratory data; and/or instrumentation/computer data may be in conflict with other sources of data.

4. Modern day plants have a great deal of variable interaction. This results in difficulty in isolating the real problem affecting either independent variables or strong correlations between dependent variables. While a strong correlation between dependent variables may be of interest, it rarely results in the solution to problems. An independent variable can be changed or set by an operator or by operating directives. Dependent variables are those that are changed by the reaction of the process. In the plastic pellet example given earlier, the independent variable is the production rate and the dependent variable is the fraction of particles with voids.

5. Besides a high degree of variable interaction, there is also a high degree of interaction between various engineering disciplines. Thus, what appears to be an obvious mechanical engineering problem often has its true roots in chemistry and/or chemical engineering. The converse is also true. This confusing scenario often leaves the process operator caught in the middle, not knowing which course of action to pursue.

6. System dynamics involve long holdup times. In the modern day process, there is usually an incentive to push the process to higher efficiency or higher purity. This usually leads to longer residence times in equipment. Problem solving with long residence time equipment requires the use of a dynamic model. Unfortunately, when faced with the need for a dynamic model, the problem solver will often take one of two unsatisfactory approaches. He will give up on the basics and say, "It's too complicated." Since the dynamic model is truly required to solve the problem, the problem solver must now take an approach that can be characterized as "guess work." The other extreme is that he will begin the development of an elaborate, technically correct model that will probably not be finished in time to be of any assistance. Both of these approaches overlook the fact that there are ways to build simple, technically correct dynamic models. These simple models will contain assumptions, however, these assumptions will still provide a model with sufficient accuracy to solve industrial problems.

7. Engineering principles are often inadequately applied by operators as well as engineers. In today's industrial environment, pressures to perform at a minimum cost and manpower commitment often encourage "shooting from the hip" as a problem-solving technique. This may be completely appropriate in some limited situations. However, the purpose of this book is to address the chronic problem that is only wounded by the "shoot from the hip" technique. The modern chemical engineering curriculum, while providing an excellent theoretical foundation, often fails to adequately stress the application of fundamentals. For example, Bernoulli's theorem can be used to explain inaccurate values given by the poorly designed level instrument shown in Figure 2-2. This design may have its origins in an engineering contractor or an

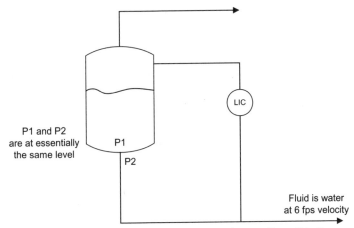

Connecting the level instrument in the process line as shown will result in the measured level reading being 0.5 ft lower than actual.

This is based on Bernoulli's theorem

$$dP/D + dV^2/2g + dZ = 0$$

where

dP = difference in pressure
D = density of liquid
dV^2 = difference in liquid velocities squared
g = gravitational constant
dZ = difference in liquid height h

At base level the pressure at the level instrument will be less than the same pressure in the drum as follows:

$$(P_2 - P_1)/62.4 + (36 - 0)/64.4 = 0,$$
$$P_1 - P_2 = 34.9 \text{ lbs/ft}^2.$$

This is equivalent to 0.5 ft in measurement of level. This ignores the friction loss in the line and nozzle.

Figure 2-2   Example of improper level instrumentation.

operator who had to improvise to get a level instrument installed in an operating plant. Either way, it must be recognized that the design will not provide accurate level readings.

8. There is often failure to use a methodical approach. While this limitation is closely allied with the previous one, it points out a need to structure even the best application of engineering principles. This structuring step is necessary to allow one to define which of the engineering principles are most appropriate. The failure to use a methodical approach could lead one to hypothesize erroneously that a fractionating tower had a plugged tray and that that was the cause of a high-pressure drop. In fact, the problem might well be associated with a change in internal vapor and liquid loading, buildup of an impurity that boils between the light key and heavy key, foaming caused by a trace impurity, or improper assumptions regarding the tower's loading point.

9. The whole picture is often not seen. The problem solver who fails to use a methodical approach is vulnerable to arriving at the wrong answer

because he fails to see the whole problem. There are often verbal clues which can hint that the problem solver is failing to see the whole picture. Some of these clues are comments such as, "That's a mechanical problem," or, "The laboratory is wrong again." While these statements may be valid, they are often indications that the problem solver is excluding essential pieces of data. It should be noted that someone using the methodical approach is less vulnerable, but still subject, to this limitation.

10. There is often an over-dependence on history. While a historical database is a mandatory prerequisite for successful problem solving, the database should be used to define deviations rather than a repository of answers. The statement, "The last time that this happened, it was due to ..." must always be tested by data analysis.

As described earlier, Figure 2-1 shows a typical industrial problem. Several of the limitations discussed above are apparent. The problem was certainly complex in that it could be caused by conditions in either the polymerization or the extrusion processes. There appears to be both a lack of a methodical approach and an inadequate application of engineering principles. In addition, while only a limited amount of data is present in Figure 2-1, the problem solution appears to be only historically based. There is no evidence that a hypothesis was developed and tested with a plant test. Was the problem solver seeing the entire picture? For example, was the independent variable polymerization production rate or extrusion rate? Was the independent variable production rate or residence time (the inverse of production rate)? Perhaps the confusion of the problem solver is illustrated by the figure, which shows the voids on the x-axis normally reserved for the independent variable.

3

SUCCESSFUL PLANT PROBLEM SOLVING

3.1 INTRODUCTION

Before beginning a discussion on how one conducts successful engineering problem solving, perhaps a definition of the activity is appropriate. Engineering problem solving is defined as the application of *engineering principles* to allow *discovery, definition,* and *solution* of plant operating problems in an expedient and complete fashion. The *discovery* and *definition* phases of problem solving are often ignored or considered obvious or unimportant. However, these phases prevent small problems from growing into large problems and allow the problem-solving phases to be done in an expedient fashion. Finding the problem involves sorting through the mass of laboratory and process data to uncover deviations that may only be a slight departure from normal, but which have the potential to grow into large deviations. Defining the problem involves developing a quantitative description of the problem specifications.

Successful engineering problem solving will always involve the following:

- A *daily* monitoring system.
- A *disciplined* (not intuitive), *learned* (not inherited) engineering problem-solving approach.
- The ability to distinguish between problems requiring technical problem solving and those only requiring an expedient answer. The ability to determine how detailed a technical analysis should be is also required to

Problem Solving for Process Operators and Specialists, First Edition. Joseph M. Bonem.
© 2011 John Wiley & Sons, Inc. Published 2011 by John Wiley & Sons Inc.

efficiently solve plant process problems. This is later referred to as *optimum technical depth.*

3.2 FINDING PROBLEMS WITH A DAILY MONITORING SYSTEM

In order to successfully find and define problems, the problem solver must obtain and maintain a historical database. The database can be maintained by using several different sources. The *managerial objective* will also be important. The managerial objective is defined as the goal that management has defined for the particular process. This goal will vary depending on the age of the process, staffing of the location and the value added by the process to name a few. Table 3-1 shows a grid of both managerial objectives and sources of data.

As an example for the use of this table, assume that a well established process is producing a commodity chemical. As a general rule, a low value is added to commodity chemicals. That is, the difference between the product revenue and the cost of production is very small. Management might elect to staff this operation so that the organization could only respond to established significant problems. Thus the managerial objectives might be characterized as Minimizing Routine Work and Maximizing Variable Retention. In this case, the number of process variables to be retained would be maximized. As shown in Table 3-1, Computer Data Storage would be the desired source of data to fit this objective. If a problem developed, the problem solver could then go back and use the stored data to attempt to resolve the problem. He might find this difficult due to the vast amount of data that must be analyzed. In addition, the data sources

Table 3-1   Sources of historical data

	Managerial Objective				
	Minimize	Maximize Finding	Maximize Trend	Maximize Variable Retention	
Source	Routine Work	Hidden Problems	Spotting	Volume	Key[a]
Computer data storage	X			X	
Computer or hand graphs			X		X
Delta data graphs[b]		X	X		X
Communication with hourly workers		X			
Visual observation of field equipment		X			

[a]The concept of "key variable retention" involves retaining the graphs or delta data graphs of only the key variables, whereas "volume retention" involves a data source that relies on maintaining values of every variable.

[b]"Delta data graphs" are the difference between actual values and a theoretical or established value. An example of such a plot is shown in Fig. 3-1.

entitled Communication with Hourly Workers and Visual Observation of Field Equipment would likely not be available since people's memory might have faded and changes might have occurred in the field equipment.

On the other hand, if the process being considered is an unproven process and/or is a high value added process, management might elect the objective of Maximize Finding of Hidden Problems. In this case, the problem solver would use Delta Data Plots, Communications with Hourly Workers and Visual Observations of Field Equipment as his data sources. It is likely that the main source of historical data would be the trend graphs or delta data graphs. Of course, in this case, the computer would still be used to store all process variable data. However, it would not be the primary source of data for the problem solver. While this objective allows for finding problems quickly, it likely will require more technical and/or operations staffing.

In the two cases cited above there are implicit assumptions. In the case where the managerial objective is Minimizing Routine Work and Maximizing Variable Retention, the implicit assumption is that essentially all process problems that occur can be readily solved without a detailed problem analysis. In the case where the managerial objective is to Maximize Finding of Hidden Problems, the implied assumption is that essentially all problems will require a detailed problem analysis.

If graphs are to be used in any of the cases shown in Table 3-1, they should be drawn, reviewed, and monitored on a daily basis. To monitor the process by preparing these graphs only once a week defeats the purpose of finding problems or spotting trends.

This daily monitoring system should be designed to allow the problem solver to monitor process variables by incorporating several variables into process models that summarize the operation of each section of the process. An example of this is shown in Figure 3-1. In this figure, reaction kinetics are

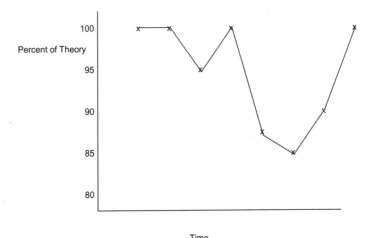

Figure 3-1   Essential variable (reactor kinetics) percent of theory vs. time.

expressed as a percentage of theoretical. Significant deviations from 100% are indicative of process impurities, catalyst contamination, or inaccurate process measurements. Thus with only a glance at the figure, it is possible to assess the status of the reactor section of the process.

Even a cursory look at Figure 3-1 will raise the question of when the problem solver declares that a problem has occurred. Is it the first drop in kinetics or the second? For each variable that is graphed, there should be a designated point at which, if the actual value exceeds or doesn't meet a certain limit, will indicate to the problem solver that a problem is likely occurring. This will be discussed later and is referred to as the concept of a "trigger point."

One source of historical data requiring elaboration is discussions with operating, mechanical, and laboratory hourly personnel. Even if the problem solver is an operator, it is likely that the observations of other operators, mechanics, or laboratory technicians will be of value. Their observations may be highly qualitative but at the same time very meaningful. For example, discussions with laboratory personnel revealed that a standard Millipore filter test used to determine the level of solids contamination in a hydrocarbon resulted in "fusing" (melting together of the two parts) of the Millipore filter container. This plastic container was known to be inert to the hydrocarbon and fusing had never been encountered before. Based on the laboratory technician's comment that the container was fusing, an investigation was initiated. This investigation showed that the hydrocarbon was contaminated with methanol. The plastic used in the Millipore apparatus was soluble in methanol. Small amounts of methanol would cause reduction of the melting point of the plastic and the fusing of the two parts of the container. The realization that the hydrocarbon was contaminated with methanol provided a strong clue for developing a hypothesis for determining the source of the known hydrocarbon contamination.

In this day and age, with multiple means of "nonpersonal" data acquisition techniques, the communications flow must be cultivated and nourished primarily by the problem solver. When it comes to cultivating communications, the best mode is face to face dialog. Telephone interactions also provide an acceptable means of communications. Written communications, including e-mail or text messaging, tend to be quick and efficient, but can often lead to misunderstandings and inaccuracies.

The observation of field equipment is accomplished by walking through the process plant and both looking at and listening to the equipment to detect any differences since the last walk through. For example, a loud noise that appears to be emanating from a process vessel might be indicative of the condensation of vapor inside of the drum. A problem solver on a walk-through may observe a new sample connection which, on closer examination, may appear to be installed in such a fashion that it will not give a representative sample. These observations by themselves may not be problems, but they are sources of data that can be considered when other problems are detected. The problem solver should make notes on anything that seems different. These notes will provide data with a time stamp that can be used for future references.

It is inadequate to only record data and collect observations. The examination of the data can best be made with "trigger points." "Trigger points" are limiting values of either laboratory analyses, instrument readings, or computed variables. If the variable being monitored is outside of these limits, the successful problem solver will declare that a problem exists and begin to solve the problem. It should be emphasized that the successful problem solver will find and define the problem well before it becomes a major problem. Finding and defining the problem are the first steps toward problem resolution. Resources may not be available to resolve the problem completely; however, management will recognize that a problem has been uncovered. The "trigger point" approach is similar to that used by the medical profession. Medical and laboratory tests such as blood pressure, cholesterol level, and hemoglobin levels are used to spot minor problems before they become major problems.

Trigger points, whether used in the medical field or in a process plant, are based on statistics. This book does not cover statistics in detail. However, to introduce the concept of trigger points, it is necessary to explain two statistical functions. For the purpose of this book, the two important statistical functions are as follows:

- *Average*: This is determined by adding up the values of a variable and dividing by the total number of values. It represents a middle value of the variable. The average can also be calculated using a spreadsheet.
- *Standard Deviation, or Sigma* (σ): This is determined by a more complicated function than the average. Fortunately, this function is also available in any spreadsheet. The spreadsheet calculates the σ using all of the values that were used for the average. It represents the range of the values. The greater the σ, the wider the range of data. This is illustrated in Figure 3-2.

As shown in Figure 3-2, two sets of data may have the same average, but widely different distributions. Figure 3-2a has a very broad distribution and thus a large σ. On the other hand, Figure 3-2b has a very narrow distribution and thus a small σ. In a process plant, this is manifested in a process with a narrow distribution by a trigger point that is close to the target operating condition.

The standard deviation is often expressed as a multiple of the calculated value such as 1, 2, or 3σ. Standard deviation is abbreviated as σ. The higher the multiple, the more data is included in the range, as shown in Table 3-2. Thus the width of the data range is a function of both the numerical value of σ and the number of standard deviations. For example, if the average value of a pressure in a process is 100 psig and the 1σ of the measured values is 5 psig, it can be concluded that 68% of the measured values will fall into a range of 95 psig to 105 psig. However, if the problem solver wants to determine the

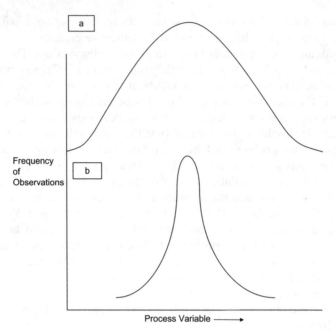

Figure 3-2 Affect of σ. (a) Large σ; (b) small σ.

Table 3-2 Percentage of values in the range

Width of Range	% of Values in Range
±1σ	68
±2σ	95
±3σ	99+

range of pressures that will include 95% of the data (2σ), he would multiply the σ (5 psi) by 2. The pressure range will be 95 ± 2σ or 95 ± 10. Thus if 95% of the data is included, the pressure range will be 90 to 110 psig.

It is clear from Table 3-2 that the multiple of σ is important in determining the percentage of the values that falls within any range. A trigger point that is based on 3σ will include essentially all of the data and cause problems to be ignored. However, if a value falls outside of this range, the problem solver can be 99+% confident that a problem is occurring. Conversely, a trigger point based on 1σ means that there is only a 68% chance that a problem is really occurring. Thus with a trigger point set at ±1σ, 32% of the announcements that a problem exists will be false alarms. As discussed later, it may be completely acceptable to have a relatively high frequency of false alarms in order to find problems at an early point.

One of the most important factors to recognize when setting trigger points is that there is a difference between using statistics to control a process and using statistics to find problems. Control statistics require that the process be greater than 3σ from target before changes are made. This implies that there is a greater than 99% confidence level that there has been a real change in the process as opposed to process variability. The successful problem solver cannot wait until he is greater than 99% confident that there is a process problem. For example, very few car owners wait until they are greater than 99% confident that they have an automobile problem before they begin a problem-solving activity.

Trigger points can be set for the following different types of variables:

- *Theoretical/Laboratory/Pilot Plant Demonstrated*: Each of these variables would have a "lumped parameter constant" (to be discussed later) that can be calculated from plant data. These constants can then be compared to similar constants demonstrated in the laboratory or pilot plant or that can be developed from theory. Examples of these are reaction rate constants (demonstrated in the laboratory or pilot plant) or fractionation tower tray efficiencies (demonstrated by theoretical calculations).

- *Plant Demonstrated*: These include variables that are equipment-related, such as production, purity, slurry concentration, or additive controllability. They can only be demonstrated in a commercial size plant where full-scale equipment is utilized.

- *Vendor Demonstrated or Guaranteed*: These will be almost exclusively equipment items. These variables will include items such as highly specialized valves, volatile removal equipment, or heat exchange equipment.

Statistical techniques can be utilized to set trigger points. For example, in a process demonstrated to have a catalyst efficiency of 5000 with a 1σ of 200, a low trigger point of 4900 would be ludicrous. Conversely, a trigger point of 4400 would cause many problems to be ignored. Obviously, determining a meaningful standard deviation is mandatory if the trigger point approach is to be utilized.

It is likely that in an industrial process, the standard deviations of essential variables are not well known. Rather than doing elaborate laboratory statistical studies, a more expedient approach involves developing approximate standard deviations and allowing the daily process monitoring to help determine the real commercial standard deviation. This approximate standard deviation can be determined by examining at least 20 values of a variable obtained while the plant under consideration is operating at steady state and calculating σ from the equation below:

$$\sigma = (V_{max} - V_{min})/6 \qquad\qquad (3\text{-}1)$$

where

V_{max} = maximum value of the variable in the data set

V_{min} = minimum value of the variable in the data set

The σ can also be calculated more exactly using the algorithms available in spreadsheets.

In a new process, there is great value in setting the standard deviation and/ or standard deviation multiple on the low side and attempting to explain as many deviations as possible. The tightening of the standard deviation for a new process will cause the maximum number of problems to be uncovered while management's attention is focused on getting the new process operational and adequate resources are available to solve problems. The opposite approach, that is, having a large standard deviation and/or multiple, will result in an apparent good startup followed by a multitude of problems 6 to 12 months later. These problems that occur in 6 to 12 months were actually present during the startup as small problems that went undetected due to the large standard deviation and/or multiple being utilized.

While problem solvers generally think in terms of negative deviations (failure to achieve a target), positive deviations must also be considered. For example, a critical heat exchanger that had a known heat transfer coefficient of $120 \pm 10\,BTU/hr\text{-}ft^2\text{-}F$ suddenly began operating with a coefficient of 150. An investigation is warranted to determine what had happened to cause an apparent new base line. This investigation of positive deviation will often lead to new or improved operating procedures.

While the actual setting of trigger points depends on the process as well as the individual company, Table 3-3 shows some suggested trigger points. It should be recognized that each of these is based on a statistical approach once a standard deviation has been developed or approximated.

Table 3-3   Suggested trigger points

Magnitude of Variable on Profits	Trigger Point	Probability of Type 1 Error (%)[a]
Very significant (or new process)[b]	1σ	32
Moderate	2σ	5
Insignificant	3σ	<1

[a] A Type 1 Error is the probability that a problem would be declared when no problem really existed. For example, if the trigger point criterion is set at 1σ, there is a 32% probability that a declared problem is really just a normal fluctuation in the process. However, there is a 68% probability that a real problem exists.

[b] Items that could cause a very serious upset (e.g., plant shutdown) should be evaluated using a one sided test against a criterion of being 60% or less sure that you are right. Thus, a trigger point of only 0.3σ could be important.

In summary, the key concepts in the use and the definition of trigger points are as follows:

- They should be based on statistics and theoretical values when possible.
- The criterion for declaring that a problem exists is different from the criterion for taking control action.
- The criterion for declaring that a problem exists will be a function of severity of the problem. In addition, the point on the learning curve for specific processes should be considered.
- Positive deviations must always be considered.

Another concern involved with the operation of a full-scale commercial unit is that some problems can be caused by transient process upsets. An adequate explanation of these upsets will usually require extrapolation to steady state condition. For example, an impurity is present for only 30 min in the feed to a reactor with a residence time of 3 hr, and causes the conversion to drop 2%. Determination of the seriousness of this feed impurity will require extrapolation to steady state conditions. The approach to developing a simplified dynamic model is discussed in Chapter 10.

While the daily monitoring system has been discussed primarily in process engineering terms, it can also be used for following mechanical equipment. One of the essential areas that can be monitored is "mean time between failures." This is the time that a piece of equipment is in service before it fails. Well-kept records will allow operators to determine whether there is any change in the failure history of a piece of mechanical equipment. In many process plants there is a strong relationship between the process and the mechanical equipment, so the problem solver should be careful that he does not exclude events that are occurring in the plant because they are not strictly in his area of training or specialization. For example, a decrease in mean time between failures of a mechanical seal may be related to the presence of very small particles in the seal flush fluid. The presence of these particles may be related to some change in process conditions.

The implementation of an effective daily monitoring program can be established using the information discussed above along with the following guidelines.

1. Pick 6 to 10 essential variables and graph then (by computer or hand) on a continuous daily basis using delta graphs and theoretically determined target values. Combine as many variables as is justified based on theory into single graphs. For example, the graph shown in Figure 3-1 combines such variables as catalyst efficiency, reactor residence time, production rate, reactor temperature, and reactor pressure into one graph.
2. Establish positive and negative trigger points for each variable. Compare the actual value to the trigger points on a daily basis.

3. On a daily basis, either obtain comments from others or observe the process and follow up on any unusual comments or observations.
4. Visually observe all equipment in the field at least weekly.
5. Store the essential variable plots so that this information can be easily accessed.

3.3 SOLVING PROBLEMS WITH A DISCIPLINED AND LEARNED PROBLEM-SOLVING APPROACH

A disciplined and learned problem-solving approach is a technique that allows one to determine if the problem really occurred, specify the problem in quantitative terms, and resolve the problem accurately and quickly. The approach discussed here differs significantly from techniques discussed in traditional problem-solving courses. The approach discussed in this book emphasizes using techniques that will verify whether the problem really occurred. Many problems presented either are not real problems or are radically different from the way they were first described.

In addition, this book emphasizes the need to use engineering principles when formulating a hypothesis to explain the problem. In the void problem described earlier, the relationship between voids and production rate is an idea or vision. A scientifically correct hypothesis would be developed by exploring the following logic path along with appropriate calculations. This logic path is as follows:

1. Voids are caused by immiscible volatiles.
2. These volatiles are present due to either or a combination of excessive immiscible volatiles in the feed or from a steam leak, poor mass transfer, and/or lack of residence time in the dryer.
3. At this point additional data could be collected and hypotheses could be developed that explain the data and observations associated with the voids on this specific grade.

The approach in this book also emphasizes that any hypothesis must be confirmed with a plant test, through calculations, or by making "directionally correct changes." A successful plant test is one that conclusively proves or disproves the hypothesis. The concept of confirming a hypothesis by making directionally correct changes will be discussed later. The approach in this book emphasizes that a problem solution must not create new problems.

The *Disciplined, Learned Problem-Solving Approach* consists of the following five steps:

Step 1: Verify that the problem actually occurred. Communications in an operating environment are almost always second- or third-hand

and are often highly garbled. The problem solver must have a means to reduce the confusion at this point.

Step 2: **Write out an accurate statement of what problem you are trying to solve.**

Answers to the following questions may be helpful:

- What happened?
- When did it happen?
- Where did it happen?
- What was the magnitude of the problem?
- What else happened at the same time or shortly before?
- What actions are you planning?

Step 3: **Develop a theoretically sound working hypothesis that explains as many specifications of the problem as possible.**

Step 4: **Provide a mechanism to test the hypothesis.**

Step 5: **Recommend remedial action to eliminate the problem without creating another problem.**

The problem verification phase may be the most overlooked part of this five-step procedure. The problem often arrives at the problem solver so jumbled that the best approach is to go directly to the "horse's mouth." For example, by talking to the operator who is having an equipment-related problem or to the laboratory technician who got a strange result, the problem solver can find out exactly what was observed. He will often find that the real problem is considerably different than what was described in an e-mail that he received. Problem verification may also take the form of data verification. While this is the subject of a later discussion, it should be noted here that application of engineering principles can often eliminate a problem by determining whether the alleged problem was caused only by a defective instrument. For example, an engineer sent to investigate the poor operation of a 40 psig steam desuperheater found that the measured steam temperature was below the temperature of 40 psig saturated steam and yet water did not appear to be present in the steam. Specifically, the measured temperature of the 40 psig steam was 280°F. The boiling point of water at 40 psig is about 286°F. Since this is a theoretically impossible situation, he began to investigate the accuracy of the instrumentation. He determined that the steam temperature instrument had been incorrectly calibrated.

The person directly involved with the problem can usually be helpful in the problem specification phase (step 2). However, his knowledge base may not allow him to formulate technically sound hypotheses, although he would normally do so. At this point, it is important to focus on the activities of step 2 (writing an accurate specification of the problem). While this description does not have to be a formal document, shortcuts or even shorthand meant to facilitate a quick answer will be counterproductive. The problem statement

Table 3-4   Problem specification example

SHORT TITLE OF PROBLEM_____

DESCRIPTION OF EVENT (make sure that step 2 is utilized to provide a complete problem description)_____

_____

_____

_____

_____

_____

HOW THE PROBLEM WAS DISCOVERED (was it by data plotting, operator discussion, etc.)_____

_____

PRELIMINARY PROBLEM ASSESSMENT
  COST OF PROBLEM (HIGH, MODERATE, LOW)_____
  IS IT AN OPERATING OR TECHNICAL PROBLEM_____
  IS THERE AN OBVIOUS IMMEDIATE FIX_____
  IF YES, WHAT IS PROBABILITY OF SUCCESS_____
  IF NO, WHAT AMOUNT OF EFFORT IS INVOLVED
  IN PROVIDING A FIX?_____
  ARE YOU ACTIVELY WORKING ON THIS PROBLEM_____

should be as short as possible while still including pertinent data. There is great value in writing out the problem specification using a structured approach. The structured approach provides a means to uncover gaps in the data. In addition, the writing process forces one to clarify data and thought processes.

Table 3-4 shows an example of a problem statement format that could be used. The key part of this format is the problem statement (i.e., the description of the event). The other parts of this format may or may not be of value depending on the organization needs.

The purpose of this format is to provide a simplified communication tool between the problem solver and different managerial layers, and to provide a format to allow the problem solver to both state the problem in problem-solving terms and assess the severity and solution difficulty of the problem.

While this form is only presented to serve as an example, there are two important concepts involved in using this or similar forms. The form should be kept as simple as possible. In addition, the tendency of management to review and edit all documents must be avoided. It should be remembered that this form is only a device to advise management of the status of problem-solving activity in the problem solver's realm of responsibility. The involvement of bureaucracy or any type of editing will be counterproductive and will often reduce the desire of the problem solver to use this technique.

The development of a theoretically sound hypothesis (step 3) to explain the problem is an essential concept in allowing industrial problems to be eliminated. A cause-effect relationship does little good unless the cause can be eliminated or understood. For example, in the void problem discussed earlier, reduction of the production rate only masks the problem rather than eliminating the problem. An example of a theoretically sound hypothesis for this problem is as follows:

> There is a condensate leak causing water to flow from the steam side of the indirect dryer to the polymer side. This water is trapped in the pores of the polymer flakes. It is not removed when it is heated in the extruder because the extruder does not have a vent. As the particle is cooled, the water condenses, forming a second phase in the polymer particle. This second phase is what causes the discontinuity in appearance.

This hypothesis must be tested against plant data, but it could explain both the appearance of voids and the sensitivity to rate. As the production rates and heat input requirements are increased, the steam pressure on the dryer would have to increase in order to provide the temperature-driving force necessary to provide more heat input. This increase in steam pressure would create more leakage potential.

Chapter 6 provides more information on how hypotheses can be formulated. The development of theoretically correct hypotheses will involve the application of engineering principles. Some of these applications are described in the following paragraphs.

Unit operations and/or equipment design calculations can be used to formulate hypotheses associated with pump or compressor motor overloading. For example, changes in the pump or compressor horsepower requirements that occur as the composition changes might be used to determine why a motor overloads. Another example is that the calculation of the amount of condensate produced from a steam turbine might be used to show that a steam trap was being overloaded, resulting in the poor performance of a heat exchanger.

Unsteady state accumulation calculations that allow analysis of a process in a dynamic mode could be used to determine how fast propane builds up in a polypropylene process. These calculations could also be used to determine how many displacements of a system are required to achieve a given degree of cleanliness during a transition or startup operation.

Mass and energy balances could be used to analyze steady state or dynamic operations. Examples of the use of these balances for dynamic operations are:

- How hot would the wall of a reactor become if heat transfer failed?
- How long could a process operate without cooling water?

The development of a theoretically correct working hypothesis is mandatory to reduce the unlimited number of hypotheses to the few that make sense. Problem solving that is not based on theoretically sound hypotheses will degenerate into unstructured brainstorming. Unstructured brainstorming quickly becomes a contest to determine who can generate the greatest number of hypotheses (sound or unsound).

As the next step (step 4) is addressed, the definition of a successful plant test must be considered. A successful hypothesis test is often thought of as one that proves the hypothesis is correct. However, disproving a proposed hypothesis is as valuable as proving one. Therefore, the definition of a successful hypothesis test is a test that either proves or disproves the proposed hypothesis conclusively. A failed hypothesis test is simply one that is inconclusive. Whether the test proves or disproves the hypothesis, the results of the test must be documented. Even for a test that disproves the hypothesis, documentation is important. This will avoid any chance of repeating the test later. This subject is covered in greater detail in Chapter 12.

The mechanism to test the hypothesis can consist of a plant test of new operating conditions, an increase in data collection frequency and/or new data, a series of calculations, or a temporary mechanical fix.

Regardless of which mechanism is selected to test the hypothesis, a great deal of salesmanship will be required to obtain the necessary cooperation from all parties that are involved. The first meeting in which the hypothesis test is proposed may be the problem solver's first encounter with the individual who originally uncovered the problem. Regardless of whether this is true or not, the carefully prepared problem statement and a statement of the theoretically correct working hypothesis will be very beneficial at this point. These two documents, along with the proposed hypothesis test, will provide an outline of the following:

- What problem are you trying to solve?
- What is the working hypothesis?
- How do you plan to prove the hypothesis?

The mode to a successful hypothesis test often lies in the hands of the hourly personnel. If the hypothesis is to be demonstrated by a plant test or by any technique that involves the hourly work force, the need to communicate the goals of the test must not be overlooked. This will also be an opportunity to explain the theoretically correct working hypothesis. This pretest communication is an excellent opportunity to teach and train as well as to obtain support for the test. A test that fails because, allegedly, "The operator did not want it to succeed," usually indicates that there was inadequate communication with the operator. The successful problem solver will always be backed by the hourly work force, who also want the test to be successful. Post-test communication is also of value. Such items as the test results, the conclusions, and future plans will help ensure future positive results.

While a plant test is a typical approach to testing a hypothesis, increased data collection can also be used as a test mode. While this appears to be an obvious statement, the existence of highly specialized techniques for obtaining additional data should be considered. Examples are as follows:

- Temperature measurements using infrared detectors can be used to either supplement and/or confirm existing instrumentation.
- High-speed data acquisition devices can often be of benefit in determining the exact sequence of events.
- A specialty designed venturi flow meter can be used to detect the presence of two-way flow in a pipe.
- X-ray pictures of equipment can be used to confirm the presence of a plugged downcomer or a damaged fractionation tray.
- Qualitative laboratory tests can be used to confirm the presence of an element that could only be present if an O-ring were failing.

The important considerations are to use all the resources available and to think outside the box to allow the proposed hypothesis to be conclusively tested.

While it is similar to a plant test, a temporary mechanical fix can be used to provide a test of a hypothesis. This approach provides a circuitous route to proving or disproving a hypothesis. In cases where a plant test is undesirable or would require an excessive amount of time, a temporary mechanical fix may allow confirmation of the hypothesis. The problem solver will use logic prior to the mechanical fix to specify what criteria would be required to demonstrate that the hypothesis was correct. The logic might be stated as "If the hypothesis is true and the proposed mechanical fix is made, the following will be observed _____." If the anticipated results are obtained, the hypothesis is confirmed with some degree of certainty. However, there is always the possibility that the hypothesis was wrong and the mechanical fix, while providing the anticipated results, did so because of a different hypothesis than the one proposed. A specific example of this approach is discussed in Chapter 4.

Once a proposed hypothesis has been demonstrated to be true, the problem must now be eradicated (step 5). The three keys to step 5 (recommend remedial action to eliminate the problem without creating another problem) are as follows:

1. In order to avoid creating another problem with the solution to the initial problem, a thorough potential problem analysis should be conducted. A potential problem analysis is a technique for visualizing what problems may occur if the recommended solution is implemented. Once a potential problem is discovered, consideration can be given to eliminating or ameliorating the problem via preventative or contingency actions. This analysis should also include safety aspects.

2. Make sure that the problem solution is the simplest one that will work. Keeping things simple is even more important in problem solving than in plant design. Make sure that, in attempting to provide a perfect solution, the solution's complexity does not create a trap.
3. Make allowances for follow-up. New operating or maintenance techniques will require a great deal of nagging, hand holding, and coddling to keep them from being forgotten or ignored.

3.4   DETERMINING THE OPTIMUM TECHNICAL DEPTH

Any discussion of optimum technical depth is meant to apply to those process problems that require serious engineering considerations. There are many process plant problems that are best solved by intuitive judgment and experience-based know-how as discussed in Chapter 1. Some of these manifest themselves in emergency and startup situations. In those situations, there is no question of optimum technical depth. Things must be done quickly with little time for introspective analysis. The concept of technical depth does not mean calculations or analyses performed by a graduate engineer, but rather calculations or analyses performed by an operator or process specialist with some training in technical calculations.

For those problems that require a more in depth analysis, there will always be a question of the required technical depth. For the purpose of this discussion, *optimum technical depth* can be defined as "the ability to compromise between expediency and thoroughness in order to solve a process problem in a minimum amount of time." This definition is shown schematically in Figure 3-3.

This is an exceptionally difficult area to quantify. It will vary greatly from company to company. Even in the same company different divisions appear to have different standards. A few definitions are required before proceeding further:

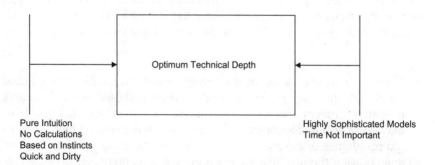

Optimum Technical Depth

Pure Intuition
No Calculations
Based on Instincts
Quick and Dirty

Highly Sophisticated Models
Time Not Important

Figure 3-3   Optimum technical depth schematic.

Confidence level is defined as the probability that the recommended problem solution will completely eliminate the problem without creating another problem. There are two different confidence levels to be considered. One is the required confidence level as suggested by management. The other is the probable confidence level as assessed by the problem solver. In order to avoid misunderstandings, the probable confidence level should be greater than the required confidence level.

Project execution time is the amount of time required from the time that the problem solver begins to work on the problem until the problem is solved. This involves time for data collection, data analysis, and implementation of operating changes or installation of mechanical equipment. Obviously this can be as short as a few days to as long as several months.

In spite of the difficulty in quantifying optimum technical depth, the problem solver should give consideration to this variable prior to initiating problem-solving efforts. Some suggested guidelines that may help in quantifying the optimum technical depth are provided in the following paragraphs.

The probable confidence level that the problem solution is correct is directly proportional to the technical depth involved in the problem-solving activity. For example, a pressure drop calculation that assumes the length of the line is about 200 ft is much less accurate than a calculation based on line measurement and a count of the number of fittings.

The required confidence level in an industrial environment is much lower than that in an academic or research environment. Courtrooms are filled with examples of alleged inadequate required confidence levels within the pharmaceutical and medical research fields. In an industrial environment where product liability is not an issue, the daily cost of the process problem often dictates the need for a lower required confidence level. The exception to this is where safety or product liability is involved. In these cases, there is a need to have a high degree of confidence that the urgency to solve a process problem does not create a product liability or safety-related problem.

The required confidence level is directly proportional to the cost and/or the execution time of the solution. Often, the solution to a process problem involves the engineering and construction of additional facilities. This can require a period of 12 to 48 months, depending on the complexity of the design. These facilities can often be very costly. If the chosen problem solution will require additional facilities, the problem solver should have a great deal of confidence that the revisions will result in a true problem solution. On the other hand, there are problem solutions that require minimal cost and can be installed quickly. These will require a lower degree of confidence prior to installation.

The required confidence level is also directly proportional to the cost of the problem, that is, the required confidence levels for solutions to costly problems are higher than those for less costly problems. In an industrial environment, costly problems also get the greatest visibility. That is, they get more management attention and, as such, require a higher degree of confidence in the problem solution. The less costly problems that require a long execution time or involve a large expense for equipment also require a high confidence level in the chosen solution. The less costly problems that can be solved by a quick, low-cost fix do not require as high a confidence level.

Unfortunately, the very expensive problems often require a detailed technical analysis. Since they are expensive, they place a great deal of pressure on the problem solver to develop a quick fix. Rather than doing the required technical analysis, the problem solver often submits to the temptation to "try something." He then finds himself spending some of his limited amount of time implementing the "something" multiple times rather than doing a detailed technical analysis.

Another aspect of assessing the optimum technical depth involves estimating the project execution time. For typical engineering projects, most industrial companies have well-established project execution times. However, for problem solving, these engineering and construction guidelines will likely not be applicable. The problem solver is the best equipped person to make an estimate of the work that needs to be done and the amount of time and resources that will be required. Once the number of man-days for completion of the project is known, the manning can be estimated. As a general rule, if the estimated project execution time exceeds 3 months, it will be desirable to increase the manning so that the execution time can be reduced to 3 months or less.

There is always a minimum degree of confidence that is acceptable regardless of the cost of the problem, the cost of project execution, or the length of project execution. It would seem that one should be at least 70% confident of the problem solution before it is proposed.

These concepts are illustrated in Figure 3-4. In this figure, the required and probable confidence levels are shown on the y-axis. The x-axis is the

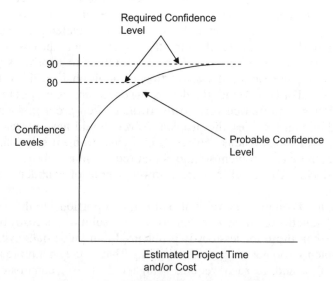

Figure 3-4   Estimated project time and/or cost.

project execution time and/or project cost. Thus as the project execution time and/or the project cost increase, the probability of success also increases. Two levels of required confidence are shown (80 and 90%). The required confidence level is set by discussions at the start of the project. It should be recognized that these discussions may start off with generalized statements such as "Just get it right," or "Do something quick." These can be translated into required confidence levels that provide some idea of what degree of certainty is required before a recommendation is presented. Obviously these are very subjective evaluations. However, the time that it takes to carry them out will eliminate future disappointment when recommendations are presented. This effort will also allow the problem solver to indicate to management whether the project can be accomplished with the required confidence level within the time and cost constraints that have been determined.

As indicated earlier, the goal of any problem-solving exercise is to obtain a true solution to the process problem in the minimum amount of time. It may be possible to do this by either:

- A detailed analysis that leads to one unique solution with a high probable confidence level that exceeds the required confidence level.
- A multitude of attempts to solve the problem. Each of these attempts will likely have a probable confidence level less than the required confidence level. However, because each of these attempts is done sequentially, the problem will eventually be solved. It should be noted that this concept still requires technical analysis to confirm that each attempt to solve the problem is a theoretically correct hypothesis. The technical analysis is not a detailed analysis and, therefore, each attempted solution has a low probable confidence level.

It is possible that the problem solver will have to consider both of these execution approaches. Management may indicate that the required resources are not available or that the required execution time is too long for the alternative with a high probable confidence level (detailed analysis). In this situation, the problem solver can help make an execution plan decision by providing management with his best assessment of the alternative execution strategy. One way of doing this is illustrated in Figure 3-5. The hypothetical example shown in this figure presents two approaches to solving the same problem. In this example, there is one unique solution to the problem. In one approach, this solution can be reached with a 90% probable confidence level with detailed study. The x-axis is the cost and/or the length of time required to reach this solution. The y-axis represents the probable and required confidence levels. In the first approach (detailed study), the required confidence level is reached by a single path. The more detailed the study is, the higher is the probability of success. In the alternative approach that uses multiple attempts to solve the problem, the required confidence level is reached, but only after several failed attempts.

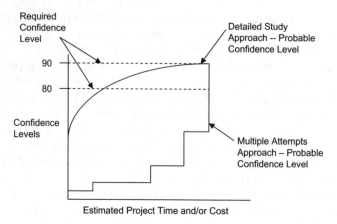

Figure 3-5   Detailed study approach compared with multiple attempts approach.

In this example, the times/costs have been adjusted so that the 90% confidence level is reached after the same expenditure of cost and/or time for both approaches. This is not the normal chain of events. Normally, the detailed study will allow you to reach the required confidence level in less time and at a lower cost. Figure 3-5 also shows what happens in the hypothetical example if the required confidence level is reduced to 80%. In this case, the detailed analysis will allow you to reach this level faster and/or at a lower cost. This is more typical of industrial problem solving. Although this is a highly theoretical example, it is given to illustrate the thought processes that should be followed in assessing the best approach to reach a final, successful problem solution in a minimum amount of time and/or cost.

Once the solution path can be agreed upon, it can be used to steward the progress of the project using the confidence level versus time/cost type of relationship that was developed to determine the project execution approach. In Figure 3-5, if the detailed study route was chosen as the project execution strategy, the problem solver should feel about 85% confident in the approach when the project is about 50% complete. This is true whether the project is building a mathematical model or installation of new equipment. Progress reports would consist of status of the project as well as an indication of probability of success. This accountability will allow for any necessary midcourse corrections.

In the second approach of multiple attempts, the relationship can also be used to steward progress. Management reports in this case will likely consist of reports of failed attempts and number of trials remaining. This reporting of multiple failed trials often leads to loss of management confidence in the problem working process.

While it may seem that the above considerations are not worth the effort required, it should be recognized that whether this type of thinking is quantified or not, it happens when any problem-solving exercise is being considered.

Proof that such thinking is present is found in statements such as, "We have to try something quickly," "That approach is nice, but it takes too long," or, "Let's just put in a treatment bed." When faced with these criticisms, the wise problem solver will attempt to define a cost/timing and confidence level analysis of the various approaches. The above technique is one approach to try to quantify this analysis.

The approach described here can be criticized as attempting to quantify items that are so subjective that they cannot be quantified. However, it should be recognized that whether these items are quantified or not they are always present in the minds of management and the problem solver. Through attempts to quantify such areas as required confidence level, probable confidence level, and execution times, better decisions on approaches and resource allocation will result.

3.5 USING THE DIRECTIONALLY CORRECT HYPOTHESIS APPROACH

There are times when the problem solver will be faced with a hypothesis that appears to be correct, but because of lack of calculation techniques or lack of technology correlations, the hypothesis cannot be proven with calculations. If the problem solver truly has a "directionally correct hypothesis" in mind, this knowledge may by itself lead to an effective problem solution. This approach differs from the trial and error approach where calculations can be made but are not because of the time or cost involved. This directionally correct hypothesis approach assumes that if one can make a small and low cost (in either time or money) change to an independent variable, the impact of this change will by itself either prove or disprove the working hypothesis. The change must be small enough so that no other parts of the process are impacted, but large enough to be confident that the impact (if any) on the dependent variable is statistically significant. As a general rule, the problem solver should have at least a 75% confidence level that the independent variable will affect the dependent variable. The sources of these directionally correct hypotheses are likely to be experience-based. The problem solver is likely to have experience with a similar process or similar piece of equipment. In his experience, a problem was solved by a change in an independent variable. He thinks that a similar change will solve the current problem. The hypothesis could be tested by making small and low-cost changes in the process. Since this small change would be expected to make only a small change in the dependent variable, it would be necessary for the process to continue operating in the new mode until enough data accumulated to statistically prove or disprove the theory. It should be noted that this method still requires technical analysis and a theoretically correct hypothesis.

An example of a situation in which this approach can be used is that of an engineer whose previous experience indicates that there should be a direct correlation between a quality attribute of a product and a plant operating variable. His experience is based on a similar but not identical process. However, in the existing process there is a large amount of scatter in the correlation of the product attribute and the plant operating variable such that he has only a 60 to 70% confidence level that there is any correlation at all. Increasing the plant operating variable is the directionally correct approach to increase the product attribute. As a test, the plant operating variable can be increased a small amount to determine if there is any change in the product attribute. Over an extended period of time, sufficient data will be collected to show whether or not there is a statistically significant difference in the quality attribute.

A similar approach to this is the need for an efficient means to test hypotheses that have been developed by the detailed analysis methods discussed earlier. It will often be desirable to test hypotheses in a low-cost fashion with minimal disruptions to plant operations. In a case study to be discussed later, two alternative approaches were available to test a hypothesis. One of the approaches would require 215 days at reduced operating conditions to provide a 90% confidence level that the hypothesis was correct. The alternative approach required no reduction in operating conditions, could be implemented immediately and was very inexpensive. Thus it met the criteria of being low cost and having minimal impact on the process. It still required 215 days to provide a 90% confidence level that the hypothesis was correct. However, during these 215 days, no reduction in operating conditions was required.

3.6   WHEN TO ASK FOR HELP

Regardless of whether one is a doctor, mechanic, or operator, there will be a time when he must evaluate his situation and his capability and knowledge and request higher level resources. This can even happen in the world of athletics. In the 1972 Super Bowl game, the Miami Dolphins' field goal kicker Garo Yepremian picked up a blocked field goal try and attempted to throw a pass. Even though he was a highly respected kicker, the pass was poorly thrown and was intercepted by the Washington Redskins' Mike Bass and returned for Washington's only touchdown of the game. If the kicker had analyzed the situation and accepted the reality that he was a kicker, not a passer, he would have simply fallen on the ball to recover it for the Dolphins. The score of the game would have been 14 to 0.

In the world of business, the problem solver may often find it necessary to request assistance from someone more knowledgeable than he is. For the process operator or specialist, this may mean requesting help from the process engineer. This request should not be considered an admission of failure, but should be considered to be good judgment on the part of the problem solver. Some guidelines for knowing when to ask for help are given in the following paragraphs.

Perhaps the most obvious time to ask for an expert's help is when a problem is being experienced with a process operation that is not covered in sufficient detail in this book. For example, this book does not specifically cover liquid-liquid extraction. The principles of this unit operation are similar to those described in Chapter 8. However, it is likely that a process operator may not be able to extrapolate information from this chapter to allow application to liquid-liquid extraction. Another area that is not covered in detail in this book is scaleup from laboratory or pilot plant data. Scaleup often creates a set of unique problems. Problems in areas such as lack of geometric similarity or mixing energy per unit volume may or may not result in scaleup problems. Both of these are examples of problems where the operator or specialist will find it of value to seek help from a process engineer.

Additional help from a process engineer might be necessary if the techniques given in this book have not resulted in a solution to a problem that has become chronic. The failure of the techniques given might be due to an improper application of either the techniques or the calculation procedures. Often, a quick review from a process engineer will allow discovery of one of the following:

- There is an error in the calculation procedure. This might be an error in the input data, such as heat content, or an error in the actual numerical manipulation.
- The calculation procedure, while being done correctly, does not apply to the specific situation. The calculation procedures in the book have limitations that are enumerated. However, at times these procedures may have been used without careful considerations of their limits.
- There is an error in the development of the problem statement, or the working hypothesis is theoretically impossible. In the previously discussed steam desuperheater problem, the original problem statement was that the desuperheater was not working because the steam was cooled too much. This was theoretically impossible since the temperature was below the boiling point of water at the measured pressure and no water was present.
- There are very subtle chemical or physical forces occurring that require the combined skills and experience of operators, chemists and various engineering disciplines to fully understand the problem. The green elastomers problem given later in Chapter 4 is an example of this.
- An adequate plant test has not been formulated. Remembering that a successful plant test is one that proves or disproves the hypothesis, the process engineer may be able to suggest an improved plant test.

The key thing to remember when additional resources are necessary is that asking for additional help is not an admission of failure, but a sign of mature judgment—don't try to be a passer if you are a kicker.

4

EXAMPLES OF PLANT PROBLEM SOLVING

4.1 INDUSTRIAL EXAMPLES

In an industrial environment where the strongest emphasis is usually placed on increased productivity, doubts about the validity of this technique will always be present. Typical questions are:

- Does this technique really work?
- On what kind of problems can it be used?
- Is it really possible in an industrial environment to use engineering calculations as opposed to intuitive problem solving?

In an attempt to answer these questions, the following examples are presented. These are all actual examples from the polymer industry. Polymer manufacturing problems are often the most difficult to solve and the author's primary experience is in this area. Two of the examples were solved successfully. The first example requires only process engineering skills and the problem solution emphasizes the need for a daily monitoring system. The second example requires minimal knowledge of statistics and mechanical engineering as well as process engineering skills. The process engineering skills required to solve these problems are covered in Chapter 5. It is likely that a process operator would require assistance when using the statistics discussed in the second example. The third example is presented to illustrate the problem of inadequate intuitive problem solving. It illustrates how a logical explanation

Problem Solving for Process Operators and Specialists, First Edition. Joseph M. Bonem.
© 2011 John Wiley & Sons, Inc. Published 2011 by John Wiley & Sons Inc.

developed by an experienced engineer can be wrong due to of the engineer's not following a disciplined problem-solving approach.

4.2 POLYMERIZATION REACTOR EXAMPLE

At 0200 hours on April 2, one of the six continuous polymerization reactors in a process plant experienced a temperature runaway. That is, the reactor temperature rose exponentially from a normal temperature of 150°F to 175°F in a 30-min period. Polymerization is an exothermic reaction that generates a significant amount of heat for each pound of polymer produced. The heat of reaction is removed by circulating cooling water. Polymerization reaction rates generally double with every 20°F increase in temperature. Doubling of the polymerization rate causes the heat generated to also double. When the reactor in question reached 175°F, the reaction was terminated by injection of a quench agent. All the other reactors were operating normally.

The temperature control system on the reactor was such that an increase in temperature caused an immediate increase in the cooling water supply flow. It was known that a small increase in catalyst rate occurred right before the temperature began increasing. However, in the past, catalyst rate increases of this magnitude only resulted in a slight temperature increase. Past experience was that following this slight increase, the reactor temperature very quickly returned to normal as the cooling water control system responded. The heat exchanger that is used to remove the heat of polymerization is periodically removed for cleaning. On April 1, the exchanger seemed to be in order.

A simplified sketch of the equipment and various data is shown in Figure 4-1. At this point, the problem solver is faced with at least three questions:

1. What should be done to return the reactor to working condition?
2. What caused the episode?
3. What can be done to prevent it from recurring in the future?

The first of these questions can be handled by a combination of good operating practices (clean out the reactor) and intuitive problem solving (the exchanger should be cleaned). However, the last two can best be approached through application of the problem-solving techniques discussed in the previous chapters.

4.3 APPLICATION OF THE DISCIPLINED PROBLEM-SOLVING APPROACH

Step 1: Verify that the problem actually occurred.

While on first glance there may not seem to be a need to perform this step, the problem solver made a cursory review of all variables to confirm that the

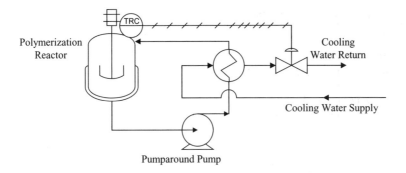

Data Values at Midnight	Technology Information
Temperatures	Reaction Heat Generated $= K\,e^{\wedge}(-11000/T)$
Cooling Water	Where K is a constant that containing monomer
In 90	concentration, catalyst concentration, reactor volume
Out 120	and heat of reaction.
Pumparound Liquid	$K = 3.9(10^{\wedge}14)$,
In 150	T is in Rankin
Out 142	The specific heat of the reaction fluid $= 0.5$ BTU/lb-F.
Flow Rates, pph	

Cooling Water: 195000
Reactor Slurry Pumparound: 2,000,000
The valve on the cooling water is 95 % open.

Figure 4-1 Reactor schematic.

reaction really was terminated due to a "temperature runaway." He found that all temperature instruments indicated an increase in temperature. In addition, the pressure on the reactor also increased.

Step 2: Write out an accurate statement of what problem you are trying to solve.

In this example, the problem that must be solved is twofold—what caused the episode? In addition, what can be done to prevent it from recurring in the future? The problem solver developed the following problem statement.

Temperature control was lost in the polymerization operation on April 2. This loss of control occurred at about 0200, following a very small increase in the reactor temperature caused by a slight increase in catalyst flow. This loss of control occurred on only one of six reactors, all of which are operating at the same charge rate on the same feedstock. The reactor had to be removed from service and cleaned prior to restarting polymerization. There was no mechanical

Table 4-1 Hypotheses conclusions

Hypothesis	Why It Can Be Eliminated
Recirculation pump stopped	"No mechanical failure"
Pumparound exchanger plugged	"No mechanical failure"
Cooling water supply lost	"No utility failure"
Catalyst activated by feedstock	"Only occurred on one reactor"
Heat generated > heat removal capability	Not eliminated

or utility failure on the reactor in question. The weather turned slightly warmer on March 30. Once the reactor temperature began increasing it rose exponentially from 150°F to 175°F in an extended period (30 min).

Determine what caused this loss of control, and once the cause has been determined, develop recommendations to prevent this problem from recurring.

Step 3: Develop a theoretically sound working hypothesis that explains the problem.

Several possible hypotheses can be proposed and the problem statement could eliminate all but one, as shown in Table 4-1. Thus a theoretically sound working hypothesis developed by the problem solver was: "The temperature runaway was caused by the fact that the rate at which heat generation increased with temperature was greater than the rate at which heat removal increased with temperature."

In order to use calculation procedures, this working hypothesis must be expressed mathematically. This can be done using differential calculus[1] as shown in equation (4-1).

$$dQ_g / dT > dQ_r / dT \tag{4-1}$$

where

dQ_g/dT = rate at which heat generation increases with temperature

dQ_r/dT = rate at which heat removal increases with temperature

This working hypothesis would predict a loss of temperature control since, as the temperature increased, the heat generation increased faster than the heat removal capability. In addition, since the rate of reaction increased with temperature, this hypothesis also predicts an exponential increase in temperature.

---

[1]While differential calculus may not be a familiar subject to a process operator, it can be easily visualized when considering driving an automobile. Acceleration is simply the rate of increasing speed as a function of time. It is called the differential of speed relative to time and is abbreviated as dV/dT, where V is velocity and T is time.

Step 4: Provide a mechanism to test the hypothesis.

While testing a hypothesis often involves experimental work, using fundamentally correct engineering calculations can also test hypotheses. In this case, experimental work would involve the risk of another loss of reactor temperature control. Thus, the problem solver used engineering calculations as the best approach to testing the hypothesis. These calculations are shown below:

Hypothesis

$$dQ_g / dT > dQ_r / dT \qquad (4\text{-}1)$$

Engineering calculations

$$Q_g = K \times e^{(-11,000/T)} \qquad (4\text{-}2)$$

where

K = a constant that depends on monomer and catalyst concentrations, reactor volume, and heat of polymerization. A typical value for this specific process and operating conditions is $3.9(10^{14})$

T = absolute temperature, °R

e = engineering constant that is equal to 2.718

A chemical engineer will recognize equation (4-2) as a typical Arrhenius equation for polymerization. The constant of 11,000 incorporates the gas constant, R. An evaluation of this equation at two different temperature levels will confirm the earlier mentioned "rule of thumb" that the rate of reaction or rate of heat generated doubles for every 20°F increase in temperature.

Equation (4-2) can be differentiated with respect to the absolute temperature, T, to yield the rate at which heat generated increases with respect to temperature as shown in equation (4-3).

$$dQ_g / dT = (K \times 11,000 / T^2) \times e^{(-11,000/T)} \qquad (4\text{-}3)$$

This differentiation is performed using concepts that a process operator may not know, but that a chemical engineer would be familiar with. If differential calculus is not used, the numerical value for dQ_g/dT can be approximated using equation (4-2) to calculate Q_g at the temperature of interest (150°F) and at another temperature slightly higher. The value of dQ_g/dT is simply the difference between the two values of Q_g, divided by the difference in the temperature.

The rate at which heat is removed from the reactor can be represented by a typical heat removal equation shown in equation (4-4). Heat balance concepts are explained in more detail in Chapter 5.

$$Q_r = U \times A \times \ln \Delta T \tag{4-4}$$

where

U = exchanger heat transfer coefficient

A = exchanger surface area

$\ln \Delta T$ = log temperature difference between the polymerization slurry side and the cooling water side

As noted in Figure 4-1, the cooling water flow is almost at a maximum (valve is 95% open), so the average water temperature will not decrease. Therefore as a first approximation:

$$\ln \Delta T = (T - T_w) \tag{4-5}$$

$$Q_r = U \times A \times (T - T_w) \tag{4-6}$$

where

$(T - T_w)$ = difference between the average reactor temperature and the average cooling water temperature

In the initial few minutes, the average cooling water temperature will remain constant. Thus differentiation of equation (4-6) gives equation (4-7). This is based on the concepts of differential calculus which state that the differential of a constant (T_w) is equal to zero.

$$dQ_r = U \times A \times dT \quad \text{or} \quad dQ_r / dT = U \times A \tag{4-7}$$

As pointed out earlier, the same numerical results can be developed by simply using two different reactor temperatures and a constant cooling water temperature. These temperatures can be used to calculate Q_r at the different temperatures and dQ_r/dT determined by dividing the difference in the Q_r values by the difference in temperatures.

By substituting actual values into equation (4-3), equations (4-8) and (4-9) can be developed.

$$dQ_g / dT = 3.9(10^{14}) \times (11{,}000 / (610)^2) \times (e^{(-11{,}000/610)}) \tag{4-8}$$

$$dQ_g / dT = 170{,}000 \text{ BTU} / \text{hr-}°R = 170{,}000 \text{ BTU} / \text{hr-}°F \tag{4-9}[1]$$

---

[1]The equality between °R and °F is valid for this expression since the temperature difference is what is being considered rather than an absolute temperature.

$U \times A$ can be estimated from the midnight values shown in Figure 4-1 using equation (4-10).

$$U \times A = Q_r / \ln \Delta T \qquad (4\text{-}10)$$

where

$Q_r = 5.75(10^6)$ BTU/hr

$\ln \Delta T = 40$

therefore

$$dQ_r / dT = U \times A = 144,000 \text{ BTU/hr-}°\text{F} \qquad (4\text{-}11)$$

As indicated earlier, the hypothesis was that "The temperature runaway was caused by the fact that the rate at which heat generation increased with temperature was greater than the rate at which heat removal increased with temperature," or, mathematically, $dQ_g/dT > dQ_r/dT$. Since the calculated value of dQ_g/dT (170,000 BTU/hr-°F) exceeds the calculated value of dQ_r/dT (144,000 BTU/hr-°F), the hypothesis was proved with calculations.

Step 5: Recommend remedial action to eliminate the problem without creating another problem.

The required remedial action developed by the problem solver consisted of providing operating procedures to ensure that the rate of heat removal always increases faster than the rate of heat generated. Mathematically this can be expressed as follows:

$$dQ_r / dT > dQ_g / dT \qquad (4\text{-}1)$$

To be conservative, a 10 to 20% safety factor should be included. Thus:

$$dQ_r / dT > 1.1 \times dQ_g / dT \qquad (4\text{-}12)$$

From equation (4-9), $dQ_g/dT = 170,000$ BTU/hr-°F. Thus:

$$dQ_r / dT > 187,000 \text{ BTU/hr-}°\text{F} \qquad (4\text{-}13)$$

$$\text{or } UA > 187,000 \text{ BTU/hr-}°\text{F} \qquad (4\text{-}14)$$

Therefore, to prevent future occurrences, he specified that the exchanger should be removed from service whenever the "UA" drops below 187,000 BTU/hr-°F.

Since UA could be easily calculated (see equation (4-10)), it became one of the key variables that was plotted and monitored on a daily basis. This could be done using the plant process control computer or by hand plotting.

Some may question the need to actually calculate a *UA* value since the narrative indicates that the cooling water was close to a maximum flow rate. While this fact should have been a red flag warning to both operations and technical personnel, there is value in being as precise as possible.

It should also be noted that the calculated value of dQ_g/dT depends on both reaction rate and reaction temperature. If the reaction rate increases (larger value of K) or the reactor temperature decreases (larger slope in the rate vs. temperature relationship), the value of dQ_g/dT will increase. This will cause the minimum value of *UA* to increase.

A potential problem analysis might reveal that the main potential problem was the degree of conservativeness used to evaluate the heat removal capacity required. A study of the variability of the rate of reaction would reveal whether 10 to 20% above the dQ_g/dT factor was sufficient. The proposed solution is certainly a simple solution. However, follow-up will be difficult because it involves requiring that operations remove a "perfectly good" exchanger from service for cleaning to avoid future episodes of temperature runaways.

4.4 LESSONS LEARNED

The value of being as quantitative as possible is actually twofold. The daily monitoring of a numerical value allows engineers to plan an exchanger down-time for cleaning as opposed to an unplanned cleaning, which will almost always occur at an inopportune time. If the heat transfer capability were followed on a daily basis in a numerical fashion, the exchanger could be removed from service for cleaning during periods of line downtimes for other mechanical reasons, or during downtimes associated with a reduction in sales volumes. In addition, the subjective observation that the cooling water flow was close to a maximum may depend on climatic conditions. These can change rapidly. Therefore, what appeared to be a situation where the exchanger had plenty of capacity changed quickly as the ambient temperature changed. If the value of the heat transfer coefficient or a comparable value had been calculated, there would be minimal affect of climatic conditions.

In situations like this, the problem solver who is under time pressures will often participate in doing whatever is necessary to get the equipment back into service. The question of what should be done to prevent the same or a similar problem from happening in the future is not considered. In this particular case, as the problem solver investigated the problem in detail, he uncovered a new area and a new technique that would prevent future temperature runaways.

While the approach that the problem solver used solved the problem and developed a system to prevent future problems, it would have been better to have a more methodical approach to developing a theoretically correct working hypothesis (step 3). The approach to developing this working hypothesis can be enhanced by a list of questions that will stimulate theoretically correct, creative thinking. This list of questions will be given in Chapter 6.

4.5 MULTIPLE ENGINEERING DISCIPLINES EXAMPLE

This example illustrates the value of a disciplined problem-solving approach when dealing with people or organizations who appear to have fixed positions based on sound logic, but inadequate data. In addition, it also illustrates the advantages of making simple changes to test hypotheses.

A process plant using a rotary filter (shown in Fig. 4-2) was plagued by excessive downtimes caused by tears of the screen cloth. A slurry of solids and liquid enters the filter at the bottom of the case. The internal drum rotates through the slurry and differential pressure forces liquid through the fine mesh screen cloth covering the rotating metal drum into the drum internal. From there the liquid and gas are removed to the filtrate handling section. The solids caught on the screen cloth are held in place by sweep gas which flows through the screen cloth into the filtrate handling section. The solids are blown off the screen by "blowback gas" as the rotating drum has gone about 270° of the total rotation. The screen cloth momentarily blows away from the rotating drum as the gas passes through the cloth. The screen cloth is held in place against the metal drum by tension rods. These retainers, while holding the cloth in place, create stress on the cloth during the blowback step. The solids that are blown off the screen are segregated from the initial slurry by a longitudinal baffle and are conveyed to the next processing step by a scroll conveyer.

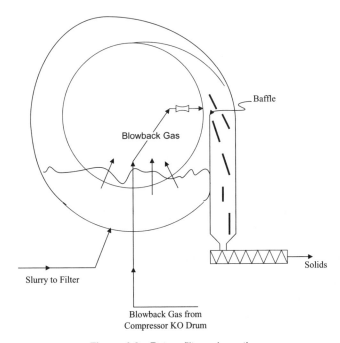

Figure 4-2 Rotary filter schematic.

The excessive screen cloth-related downtimes occurred on only one out of three rotary filters. These filters were thought to be operating under essentially the same conditions as judged by operations and technical personnel. Whenever the screen cloth would tear, solids would enter the filtrate stream, causing a shutdown of critical equipment and a resulting shutdown of the plant. After each screen cloth tear, the screen cloth and rotating metal filter drum were carefully examined. The examinations showed that the metal drum would be scratched. There was no apparent reason for the scratches, that is, there was no residual that could have caused the scratches. Solids would be present between the cloth and the drum. This was not surprising since the cloth was torn and it was known that solids had passed into the filtrate. The cloth would be torn in a circumferential manner, with most of the tears and drum scratches occurring in the middle 60 to 70% of the rotating drum.

Even after this careful observation of the filter, no consensus conclusions were reached concerning the failure. In fact, several heated arguments developed, with several fixed positions being taken. The mechanical engineers believed that the hard solid polymer particles were cutting the cloth. They believed that these polymer particles were so small, they leaked through the cloth and around the cloth-retaining facilities. The process engineer believed that some hard, metallic part of the filter was rubbing against the cloth and the metal drum. This would cause the cutting and failure of the screen cloth, letting large amounts of solids into the filtrate to scratch the metal drum. He thought that the baffle which isolated the solids from the filtrate might be the part of the filter that was rubbing against the cloth and drum. However, he had no explanation for how this might happen, since there was acceptable clearance between the baffle and rotating drum.

Since there were people in the research organization who were experienced in this process, they were also called for assistance. They believed that there was liquid in the blowback gas and that this liquid was cutting the cloth. This would allow solids to enter the filtrate and also to scratch the rotating metal drum.

Faced with such a diversity of opinions and minimal data, the problem solver approached the problem using the five-step approach discussed earlier. He made a decision to obtain as much data as possible from all sources.

4.6 APPLICATION OF DISCIPLINED PROBLEM-SOLVING APPROACH

Step 1: Verify that the problem actually occurred.

In this example, there was no doubt that the problem actually occurred. However, there was question as to whether the problem was worse than it had been in previous years. That is, problem verification consisted of considering if there had there been a change in the frequency of screen cloth tears.

Table 4-2 Mechanical history

Time Period	Mean Time between Failures (days)	Type of Tear
Past data	43	Horizontal along the tension rods that held cloth in place
Current data (all runs)	16	Circumferential
Current data excluding the very short runs	25	Circumferential

A review of mechanical records indicated the following, as shown in Table 4-2.

Obviously, a problem existed. It should be noted that without detailed mechanical records (daily monitoring), quantifying the extent of the problem would have been impossible.

A further review of what changed between the past and current data revealed that the filtration temperature on this filter was increased from 130°F to 170°F. This higher temperature was not originally considered to be a problem, as the mean times between failures on the first few runs at the higher temperature were essentially the same as they had been prior to the increase in temperature. There was a significant advantage to operating at the higher filtration temperature, so returning to the previous process conditions was not a satisfactory solution to the problem.

Step 2: Write out an accurate statement of what problem you are trying to solve.

The following problem statement was written by the problem solver: "There has been a significant increase in the screen tearing frequency that occurred on only one of three filters. This increase appeared to occur at the same time the filtration temperature was raised. In addition, to a reduction in mean time between screen failures (increased frequency), the nature of the screen failure changed. Previous failures were fatigue failures caused by the cloth being weakened during flexing while being held in place by the tension rods. The current failure is a catastrophic circumferential failure. The current failure is also characterized by scratch marks on the metal drum. Determine the cause for the significant change in screen-tearing frequency. In addition, recommendations should be made for what changes are necessary to eliminate this problem."

Step 3: Develop a theoretically sound working hypothesis that explains the problem.

Since the new failure mode appears to be related to the increase in filtration temperature, the following three hypotheses were developed.

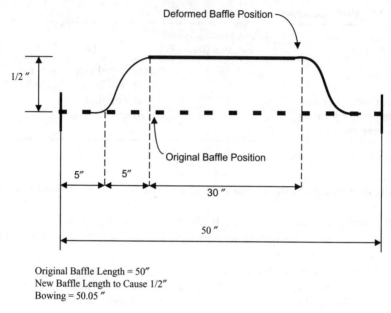

Original Baffle Length = 50″
New Baffle Length to Cause 1/2″
Bowing = 50.05 ″

Figure 4-3 Hypothetical baffle deformation, top view.

1. The screen cloth is decomposing at the higher temperatures.
2. The baffle (see Fig. 4-2) is expanding due to thermal growth and bowing into the filter cloth and metal drum.
3. The rotating drum is deforming at the higher temperatures, causing poor distribution of blowback gas. The poor distribution causes an increase in blowback gas in the middle of the drum, which then blows the filter cloth into the baffle, causing the cloth to tear.

Of these hypotheses, only the single hypothesis of baffle expansion could account for both the screen cloth tears and the scratches on the metal drum. The baffle position required to cause the observed failures is shown in Figure 4-3.

Step 4: Provide a mechanism to test the hypothesis.

This hypothesis was tested by calculations of thermal growth of the baffle. These calculations assume that the drum case will remain at the ambient temperature. The baffle, since it is immersed in the slurry, will approach the slurry temperature and experience thermal growth. The magnitude of this thermal growth will depend on the difference between ambient and filtration temperature as well as the coefficient of linear expansion. The coefficient of linear expansion for the specific metal can be found in any reference source (either handbooks or on the Internet).The growth calculations are shown below:

Given:

Original baffle length = 50 in
Original distance between baffle and rotary drum = 0.5 in
Coefficient of linear expansion = 0.000011 in/in-°F

A typical relationship relating length to temperature is as follows:

$$l_t - l_o = l_o \times 0.000011 \times dt \qquad (4\text{-}15)$$

where

l_t = the baffle length at the new temperature
l_o = the original baffle length
dt = the change in temperature, °F

Considering the baffle shape shown in Figure 4-3, the baffle would only have to grow 0.05 in to cause it to bow into the rotating drum. Thus the new baffle length would be 50.05 in. The increase in temperature that would cause this amount of growth was calculated using equation (4-15) as shown below in equation (4-16).

$$dt = 0.05/(50 \times 0.000011) = 90°F \qquad (4\text{-}16)$$

As these calculations indicated, the baffle would be expected to grow sufficiently to expand into the metal drum and screen if the differences between the filtration temperature and ambient temperature exceeded 90°F. Thus, an increase in filtration temperature from 130°F to 170°F significantly increased the probability of the baffle bowing into the drum. The fact that the baffle could also bow away from the drum without any particular consequences explained why failures did not always occur when the temperature difference approached 90°F.

Two alternatives were available to further test this hypothesis. The filtration temperature could be reduced to the level at which it had been during previous operations. A second possibility was that a mechanical constraint could be provided to cause the baffle to always bow away from the drum.

Since hot filtration was desirable for the process, reducing the filtration temperature would only be permissible for testing. In addition, the testing period would have to provide a high degree of confidence that the problem was caused by the higher filtration temperature, while taking place over a minimum amount of time.

An analysis was made to determine the minimum amount of time required to give a 90% confidence level that returning to the lower filtration

Table 4-3   Statistical data

Period	Mean Time between Failures (days)	Runs
Before hot filtration	43 ± 26	31
After hot filtration	16 ± 10	26

temperature could eliminate the problem. The basic statistical data developed for this analysis is shown in Table 4-3. In addition, the statistical approach used by the problem solver to determine how to proceed with step 4 is defined in the following paragraphs. While it is recognized that the process operator or specialist will not normally have sufficient knowledge to proceed with this statistical calculation, the calculation is shown here to illustrate the value of using statistics to determine the requirements for determining whether a process change really improved operations.

If the filtration temperature is returned to the lower value, the mean time between failures will increase to the previous value (43 ± 26). We will assume that the values of the mean and standard deviation will be the same as they were before hot filtration. The experimental test of returning to the lower filtration temperature needs to be accomplished in the minimum amount of time (i.e., with the minimum number of experimental runs). The high values of the standard deviation (26 and 10) means that more than a single test run will be required. Thus, the minimum number of runs required to prove that there has been a significant statistical improvement from hot filtration at the 90% confidence level needs to be determined. The minimum number of runs can be determined using a statistical comparison of the means of hot filtration and the results after the process is returned to the lower temperature filtration. When statistics are used to compare two values of means, a two-sided test is involved. This comparison of means will produce a numeric value such as $A \pm B$, where A is the difference in the means and B depends on the standard deviations and the actual number of runs to be used in the statistical test. To conclude that there is a real difference between hot filtration and cold filtration in the experiment, both possible values of the term $A \pm B$ must be positive. For both possible values of the algebraic manipulation to be positive, B must be less than A. Since the value of B decreases as the number of experimental runs increases, the minimum number of experimental runs will be the number that produces a positive value for both sides of the statistical test. An iterative procedure is required to develop the final answer. The final iteration is shown below:

Assumed number of runs at reduced temperature = 5

$$SE = (s_1^2/n_1 + s_2^2/n_2)^{1/2} = (100/26 + 676/5)^{1/2} = 11.79 \qquad (4\text{-}17)$$

$$1/\varphi = (1/\varphi_1) \times (s_1^2/n_1/(s_1^2/n_1 + s_2^2/n_2))^2$$
$$+ (1/\varphi_2) \times (s_2^2/n_2/(s_1^2/n_1 + s_2^2/n_2))^2 \qquad (4\text{-}18)$$
$$1/\varphi = (1/26) \times (3.85/139)^2 + (1/5) \times (135.2/139)^2 \approx 5 \qquad (4\text{-}19)$$

where

SE = standard error for comparing the two means

s = standard deviation of the two samples

n = the number of runs in each sample

φ = the degrees of freedom in each sample (number of runs)

From statistical tables for t distribution (two-sided at 90% confidence level)

$\mu = 2.01$

$\therefore$ Difference in mean time between failures = $43 - 16 \pm 2.01 \times 11.79$

$$= 3 \text{ to } 50 \text{ days} \qquad (4\text{-}20)$$

When only four runs at the lower temperature are assumed, the calculated difference in mean time between failures is –1 to 55. Therefore, four runs are not sufficient to provide conclusively significant data. However, when assuming that five runs are conducted at the lower temperature, both sides of the statistical test (3 and 50) are positive. This represents the minimum number of runs that are required. At this point, the problem solver could say with 90% confidence that returning to the lower temperature filtration would lower the frequency of screen tears. However, it would be difficult to accurately define the anticipated advantage for returning to the lower temperature due to the large standard deviation. More experimental runs would be required to narrow the range of anticipated benefits.

An estimated period of time for these five runs would be 5 × 43 or 215 days. Therefore, after 215 days at the lower temperature filtration (90% confidence level), the problem solver could say that returning to the previous temperature conditions will return the average screen cloth life to 43 days. However, this does not conclusively prove or disprove the working hypothesis. It only proves or disproves the effect of filtration temperature on the average screen cloth life. There might be another potential hypothesis that explains the problem. In addition, the test does not yield any acceptable problem solution, since it was desirable to operate at hot filtration.

The other alternative testing procedure (adding a mechanical constraint to ensure that the baffle always bows away from the drum) was easy to perform and provided a good mechanism to test the "baffle bowing" hypothesis while allowing continued operation at higher filtration temperatures. However, it involved political risks, since the addition of the mechanical constraint had "never been done this way before."

Step 5: Recommend remedial action to eliminate the problem without creating another problem.

Selecting the remedial action in this example was a strong function of how the hypothesis was tested. If the process conditions were modified to allow the system to return to the lower temperature filtration for 215 days, there would be a tendency to recommend staying at the lower temperature operation as a problem solution. Note that since this was undesirable in terms of process considerations, it would not be an acceptable recommendation.

The alternative technique of mechanically constraining the baffle so that it always bows away from the drum would provide both a testing procedure and a permanent solution. Thus, after 215 days, steps 4 and 5 could both be considered to be complete. This was the alternative that the problem solver recommended for the test and for the permanent solution. The potential problem analysis focused on how to make sure that sufficient tension would be applied to the mechanical constraint to ensure that the baffle bows away from the filter.

4.7   LESSONS LEARNED

As indicated in the problem description, the initial assumption was that all of the filters were operating at essentially the same conditions. Assumptions of this kind are almost always present in any problem-solving activity. It is only when one dedicates sufficient time to analyze data that it will be found that the initial statement of "essentially at the same conditions," or, "no process changes were made," is found to be incorrect.

Often, minor mechanical changes (such as adding a baffle brace) will provide simple solutions to complex problems. However, in this case, the potential problem analysis of the proposed remedial action missed the possibility that the mechanical constraint might fall off the baffle due to corrosion, vibration, or metal fatigue. If this happened, the device would likely go with the polymer. This did happen, causing failure of downstream equipment and some contamination of the polymer with metal. If this problem had been uncovered in a potential problem analysis, preventative action could have been taken, consisting of using a backup nut on the constraining device and/ or insuring that the bolt and clamp were made out of corrosion-resistant materials. This illustrates the importance of potential problem analyses. Often, the problem solver is so intent on moving into the execution phase of a problem solution that he does not give adequate consideration to potential problem analysis. This phase deserves as much attention as does developing a problem solution.

Besides the engineering advantage of using the disciplined problem-solving approach in this example, there is also a psychological advantage. Once a person takes a fixed position on any subject, it is almost impossible to change his or her mind without sound data. Often, it takes more sound data to change a person's opinion than it would have taken to form the person's opinion if the data had been obtained prior to the development of a working hypothesis. Any of the initial positions described by the different engineering disciplines within this example could be partially supported with logic. However, it was only after the problem solver uncovered as much data as possible and developed a hypothesis based on this data that a theoretically correct working hypothesis emerged. Normally, the application of the proposed approach will significantly narrow the hypotheses down to one or two, which can then be tested in step 4 of the five-step procedure.

4.8 A LOGICAL, INTUITIVE APPROACH FAILS

A customer complaint was received at the manufacturing location of a highly regarded supplier of a baled elastomer ($12\,in \times 28\,in \times 7\,in$). The customer alleged that he had received some green bales of the product in a recent shipment. The bales were normally a yellow color. The process for manufacturing the elastomer was about 10 years old and a similar problem had never been encountered. When confronted with this complaint, the Operating Department Head used problem-solving techniques and developed the following problem statement:

> The customer complaint has been investigated. We have not made any significant changes in our operation in 10 years except for the use of "magic markers." Our operators have started carrying "magic markers" to mark equipment that requires maintenance at the next downtime. We believe that one of these markers must have fallen from one of their pockets into the extruder. The subsequent fracture and dispersion of the material caused several bales to have a green appearance.

This intuitive, logical approach overlooked several details that a more structured procedure would have uncovered. There was no verification that the problem really occurred. It would have been valuable for the customer to send a sample of the material that he received. The problem definition was incomplete. Consideration of other questions, such as the following, would have been helpful in forming a better problem statement.

- On what shift did the problem occur?
- Did other customers notice the problem?
- How many bales were green?
- Did the problem occur on all lines?
- Did laboratory-retained samples on the problem date, previous dates, or subsequent dates show a green color?
- Were there really no operational changes?
- Were the bales green when they were boxed?

In addition to problem definition failures, the hypothesis was not tested against any type of theory. For example, how much of a green magic marker would be required to turn a single yellow bale to a green color? There was no way to estimate the concentration of green magic marker in the bale since there was no knowledge of the number of green bales. No mechanism was provided to test the hypothesis; though the hypothesis could have easily been tested by dropping a green magic marker into the extruder. This would only provide a one-sided test. If the bales did not turn green, the test would be successful in that it would prove that the hypothesis was incorrect. However, if the bales did turn green, it would be necessary to consider the hypothesis in more detail. For example, the following questions should be considered:

- How many bales turned green from a single magic marker?
- Did this number correspond to the number of green bales that the customer observed?

After the manufacturing manager wrote the customer a letter of apology for a magic marker falling into the extruder, several other complaints on the same subject were received from different customers. The continued customer complaints led to the formation of a multidiscipline problem-solving team that determined, after a lengthy investigation, that the green color was associated with an obscure change in the makeup water used for the polymer-water slurry system.

4.9   LESSONS LEARNED

This example illustrates how a failure to adequately develop a problem statement can lead to an embarrassing and faulty problem solution. If the problem statement had been fully developed, using some of the questions shown above, it is likely that the problem would have been recognized as systemic rather than as an individual isolated accident. The failure to adequately define the problem then led to the point at which only a simplified, logical approach to problem solving was used. If the problem had been recognized as a systemic problem initially, the problem-solving team would have been formed at a much earlier point in time.

This example also illustrates how what appears to be an operation to which plant operators have not made any changes can be impacted by subtle and obscure changes in utilities. The assumption that "water is water" was not true in this case.

NOMENCLATURE

A	Exchanger surface area
dQ_g/dt	Rate that heat generation increases with temperature

dQ_r/dT	Rate that heat removal increases with temperature
dt	Change in temperature, °F
K	A constant that depends on monomer and catalyst concentrations, reactor volume, and heat of polymerization
$\ln\Delta T$	Log temperature difference between the polymerization slurry side and the cooling water side
l_o	Original baffle length
l_t	Baffle length at the new temperature
n	Number of runs in each sample
s	Standard deviation of the two samples
SE	Standard error for comparing the two means
T	Absolute temperature, °R
$(T - T_w)$	Difference between the average reactor temperature and the average cooling water temperature
U	Exchanger heat transfer coefficient
φ	Degrees of freedom in each sample (number of runs)

5

FUNDAMENTALS OF CHEMICAL ENGINEERING FOR PROCESS OPERATORS

5.1 INTRODUCTION

In order to solve chronic process-operating problems, some basic understanding of chemical engineering is required. This chapter summarizes some of the basic chemical engineering skills necessary to develop solutions to chronic process problems. Mastery of this chapter will aid in the formulation of theoretically correct working hypotheses. A failure to master the concepts will cause one to propose working hypotheses that are fundamentally impossible. While the rapid development of working hypotheses may seem to be the most expedient approach, the emphasis on speed will almost always delay the development of a theoretically correct working hypothesis.

The concepts are presented in a format designed to be understood by a reader with a basic understanding of algebra and chemistry. A quick review of the different means by which to measure flow rates around a process vessel will be helpful. These different measurements are as follows:

- *Mass Rate*: This is the most helpful measurement when working with material balances as will be described later. It is simply the mass rate of material leaving or entering a vessel expressed in units such as lb/hr.
- *Volume Rate*: This is a similar measurement, expressed in volumetric units such as gallons/minute.
- *Mol Rate*: The mol rate is an empirical value that is determined by dividing the mass rate by the molecular weight of the material. It has limited

Problem Solving for Process Operators and Specialists, First Edition. Joseph M. Bonem.
© 2011 John Wiley & Sons, Inc. Published 2011 by John Wiley & Sons Inc.

value in discussions involving material or heat balances or fluid flow, but has great value in discussions of equilibrium or reaction rates.

The concepts in this chapter deal with the basic aspects of chemical engineering. They do not touch on process technology. As indicated in an earlier chapter, the specific details of the process technologies are normally presented in operating manuals or licensing packages. The basic concepts discussed in this chapter are as follows:

- *Material Balances*: Around any process vessel at steady state, the material leaving (expressed in mass per unit time, e.g., lb/hr) must equal the material entering the vessel. "Steady state" means that there is no change in the vessel inventory. If there is change in the vessel inventory, the steady state assumption will not apply because material is being accumulated or deaccumulated. It is important that the material balance be expressed in mass units rather than mol or volume units. Often, if a reaction is occurring, there will not be full closure unless mass units are used. That is because reaction often changes the density or number of mols. The same concerns are present if there is a great temperature difference between the inlet and outlet flow of a vessel. In this case, the mass balance will still be valid, but because of the change in density associated with the temperature change, a volume balance will show more or less material leaving than entering.

- *Heat Balances*: When a system is at steady state, any heat added to or removed from the process must be from either the heat of a reaction or heat added/removed by way of direct or indirect heating or cooling. An example of this is an exothermic reaction, which adds heat to the process. In this case, the heat of reaction must be removed by way of direct or indirect cooling for the system to remain at a constant temperature.

- *Fluid Flow*: The flow of fluids (gas or liquids) results in a frictional pressure drop. It is of value to know how to calculate the pressure loss.

- *Equilibrium*: All chemical engineering operations have an equilibrium limit. A reaction that has reached equilibrium will not proceed further, regardless how much time is provided.

- *Kinetics*: Kinetics deal with the speed of a chemical engineering process. They can be associated with heat transfer, reaction, solvent removal, or other unit operations.

- *Equipment Design*: This section is certainly not an exhaustive discussion of equipment design. However, since equipment design is a key factor in determining the speed of a chemical engineering process, it is important to have some knowledge of equipment design factors that influence this parameter.

Each of these concepts is discussed in the following sections.

5.2   CONSISTENCY OF UNITS AND DIMENSIONAL ANALYSIS

There is a well-worn cliché used in elementary algebra which states, "You cannot add apples and oranges." If you do add apples and oranges, you cannot call them by either specific name and have to resort to calling them "pieces of fruit." Likewise, any chemical engineering analysis requires that consistent units be utilized. In the simplest examples of this concept, lb/hr cannot be added to gal/min. The values must be converted to the same set of units. The conversion of gal/min to lb/hr is very simple; it can be done by multiplying the value in gals/min by the density in lb/gal and the number of minutes in an hour. While this appears very simple, there is often a question of whether a factor should be used as a multiplier or a divisor. Often, even highly skilled people will multiply when they should be dividing. Dimensional analysis provides a means to confirm the validity of the mathematical operation that is being used. For example, to convert a flow of gal/min to lb/hr, the dimensional check is as shown in equation (5-1) below:

$$\text{lb/hr} = \cancel{\text{gal}}/\cancel{\text{min}} \times \text{lb}/\cancel{\text{gal}} \times 60\,\cancel{\text{min}}/\text{hr} = \text{lb/hr} \qquad (5\text{-}1)$$

The gallon terms in the numerator and denominator cancel each other out, as does the minute term, leaving only the lb/hr term. This can become more difficult when more complicated relationships are involved. However, the principle of dimensional analysis is still valid. While this is a simple concept, it is unfortunately often overlooked in calculations and in stating the results of a calculation. The key ideas to remember are to state the dimensions of the calculations, to be consistent with units, and to make sure that all conversions of units are done correctly.

5.3   ABSOLUTE UNITS

While absolute units of temperature and pressure are covered in all high school chemistry courses, they are such an integral part of chemical engineering that they are reviewed in this section. The actual pressure in a vessel is that which is measured plus atmospheric pressure. This can be easily visualized by considering the measurement of the pressure in an automobile tire. The stem on the pressure gage being used is displaced by a distance that is proportional to the difference between the pressure inside the tire and the pressure outside the tire (atmospheric pressure). Thus the pressure inside the tire is actually the atmospheric pressure plus the measured pressure. This is referred to as "absolute pounds per square inch" (psia). It is this factor that is important in all calculations involving material characteristics that depend on pressure, for example, density. It is also important in determining compression ratio and in almost any calculation in chemical engineering.

In a similar fashion, absolute temperature must be used in the evaluation of such items as material density, radiant heat transfer, compressor compression ratios, and head calculations, as well as other terms. While, thermodynamically, absolute zero is defined as the temperature at which all molecular motion ceases, this does not have much pragmatic meaning. When you are using the English units of temperature, Fahrenheit or °F, the absolute temperature in Rankine units, or °R, is obtained by adding 460 to the temperature in °F.

5.4 MATERIAL BALANCES

The key concept in understanding material balances is that matter can neither be created nor destroyed. This concept is valid for every process except those dealing with atomic fusion or fission. This means that if the amount of material entering a given process or unit operation is known, the amount of material leaving is also known—it is equal to the amount of material entering. However, this concept is valid for mass units *only* at steady state. In the process industry, the quantity of material can be measured by mass units (pounds or kilograms), volume units (gallons or liters), or in mols. When dealing with material balances, only mass units should be used. A process that is operating at steady state is one in which there is no component accumulation or deaccumulation. For example, if an accumulator drum is included in the process, the level must be constant for the process to be considered to be at steady state.

Figure 5-1 shows a simple storage drum. If the fluid entering the drum is close to the ambient temperature, there will be no temperature change in the

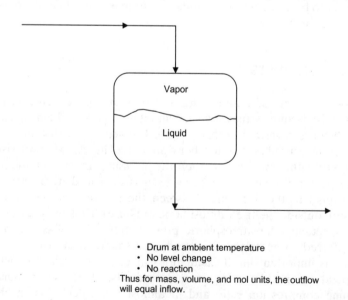

- Drum at ambient temperature
- No level change
- No reaction

Thus for mass, volume, and mol units, the outflow will equal inflow.

Figure 5-1   Material balance for simple drum.

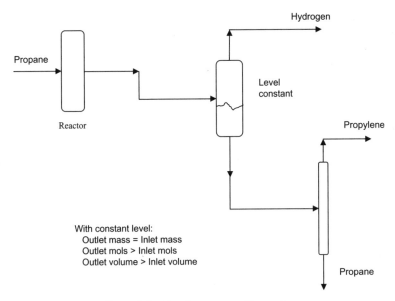

Figure 5-2 Simple process with reaction.

drum. In addition, there is no reaction occurring in the simple drum. Thus in this isolated case, the mass balance, volume balance, and mol balance will show no change. That is, the units (mass, volume, or mol) entering will equal those leaving, as long as there is no change in level.

As an example of a more complicated system, consider the reactor system shown in Figure 5-2. The reaction being conducted in the reactor is the dehydrogenation of propane to produce propylene and hydrogen. The chemistry of the process is shown below:

$$C_3H_8 \rightarrow C_3H_6 + H_2 \tag{5-2}$$

By material balance if there are 100 lb/hr of propane in the reactor feed, there must be 100 lb/hr of propylene and hydrogen leaving the reactor. If the flow meters on the feed and products validate this concept, the process is said to be "in material balance." In addition, the material balance concept can be used to confirm flow meter accuracy or determine the flow of an unmetered stream. An inspection of equation (5-2) indicates that there are more mols leaving the reactor than entering the reactor. That is because 1 mol of propane generates 1 mol of propylene plus 1 mol of hydrogen. Thus the material balance concept is not applicable for a mol balance. In a similar fashion, it can be shown that the material balance does not apply to a volume balance. In some industries such as refining that utilize volume units consistently, some processes will have either a volume gain or loss. Therefore, mass balances are the only valid use of material balances.

In addition to only using mass units for material balances, the process being considered must be at steady state. In Figure 5-2, if the level in the gas separator is constant, the process will be in material balance if the flow meters are correct. However, if the level is not constant, it is not operating at steady state and it will appear to be "out of material balance." Ensuring that a process is at steady state is imperative in any utilization of material balance concepts.

The simplified sketch shown in Figure 5-2 makes it relatively easy to understand the material balance concept. In actual practice, the process flows are always more complicated. The utilization of a technique to make sure that the input and output streams, as well as the components, are adequately accounted for is often required. One technique for this is shown in Figure 5-3. In this figure, the same process (propane dehydrogenation) is shown. A possible internal recycle stream is added. To ensure that all streams are adequately accounted for, the technique of encircling the process is shown. Using this technique, any stream that is "cut" by the encircling line must be included in the material balance. Conversely, any stream that is not "cut" is not included in the material balance. Therefore the recycle stream would not be included in the material balance.

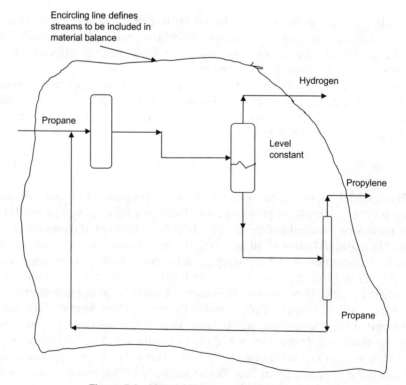

Figure 5-3 Material balance with encircling line.

The material balance concept can also be used as a technique to measure the rate of accumulation or deaccumulation in a process or process vessel. For example, if the measured flow going into and out of a process accumulator, such as the one shown in Figure 5-1, were known, the difference would represent the rate of accumulation in the vessel. For example, if the inlet flow meter to a process vessel showed 15,000 lb/hr and the outlet flow meter showed 10,000 lb/hr, the liquid would be accumulating at a rate of 5000 lb/hr or, if the liquid was water with a density of 62.4 lb/ft³, the accumulation rate would be 80 ft³/hr. As will be discussed in later chapters, this is a means of confirming the accuracy of a level instrument. However, it is also a possible source of error. If the accumulation or deaccumulation is determined as a small difference between two large numbers, a small error in either number can create a significant error in the difference.

5.5 VOLUME/MOL/MASS CONVERSIONS

As indicated, to fully utilize the concept of material balance, the rates should be given in mass units (lb/hr, kg/min, etc.). The process for converting from volume or mol units to mass units is fairly simple. If the flow rates are given in volume units (gallons, liters, etc.), they can be converted to mass units through multiplication by the appropriate density. If the rates are given in mols, they can be converted to mass units through multiplication by the molecular weight. Molecular weights and liquid densities can be found in reference books or internet sources.

However, gas densities require some elaboration. Gas densities in English units (lb/ft³) can be calculated with the equation shown below:

$$\rho = MW \times P \times 520/(379 \times T \times 14.7 \times Z) \tag{5-3}$$

where

ρ = gas density in lb/ft³

MW = average molecular weight of the gas

P = pressure of the gas in psia

T = absolute temperature in °R (460+°F)

Z = compressibility factor (will be equal to 1 for low pressures)

The basis for equation (5-3) is that 1 mole of any gas at atmospheric pressure (14.7 psia) and 520°R occupies a volume of 379 ft³. Then, by definition, 1 mole contains a mass equivalent to the molecular weight of the gas. As the gas pressure is increased or the gas temperature is decreased, the gas becomes denser, as more molecules can now fit into the same space. At moderate and high pressures, the simple ratios between temperature and pressure are not completely adequate to calculate the gas density. The compressibility factor is necessary at the higher pressures, to account for the fact that more gas can be compressed into the given volume than is explained simply by the pressure and temperature relationship.

EXAMPLE PROBLEM 5-1

Assume that you need to find the density of a mixture of propane and butane gas at the following conditions:

- The gas is 50 mol% of each compound.
- The pressure is 25 psig.
- The temperature is 90°F.
- The molecular weight of propane is 44 and the molecular weight of butane is 58.
- Since the gas is at low pressure, the compressibility is equal to 1.

When dealing with gases, a key concept to understand is that the composition given in mol % is equal to the composition in volume %. Using this concept, there are actually two different approaches to calculating the density of the gas mixture.

Approach 1: Using Mol %.

Step 1: Determine the average molecular weight of the gas. This can be accomplished with the following equation:

$$MW = (44 \times 50 + 58 \times 50)/100 = 51 \qquad (5\text{-}4)$$

Step 2: Determine the density, using equation (5-3).

$$\rho = 51 \times (25 + 14.7) \times 520/(379 \times (460 + 90) \times 14.7 \times 1)$$
$$= 0.34 \ \text{lb/ft}^3 \qquad (5\text{-}5)$$

Approach 2: Using volume % and individual gas densities.

Step 1: Determine the density of each gas using equation (5-3).

For propane

$$\rho = 44 \times (25 + 14.7) \times 520/(379 \times (460 + 90) \times 14.7 \times 1)$$
$$= 0.30 \ \text{lb/ft}^3 \qquad (5\text{-}6)$$

For butane

$$\rho = 58 \times (25 + 14.7) \times 520/(379 \times (460 + 90) \times 14.7 \times 1)$$
$$= 0.39 \ \text{lb/ft}^3 \qquad (5\text{-}7)$$

Step 2: Determine the average density. Remember that volume% and mol% are equal. Thus, the average gas density is:

$$\rho = (0.30 \times 50 + 0.39 \times 50)/100 = 0.34 \text{ lb/ft}^3 \qquad (5\text{-}8)$$

Either technique gives the same value for the average gas density.

One detail to be especially careful of when converting from one type of composition to another (e.g., volume% to weight%) is the method of using the given composition. The best method is to convert the old measurements to the new ones and to then make the calculations of average density or average molecular weight or composition. For example, the average density of two liquids cannot determined by multiplication of the densities by weight% of each component, but must be determined as shown below:

$$\rho = 100/(w_1/\rho_1 + w_2/\rho_2) \qquad (5\text{-}9)$$

where

ρ = the average density or densities of liquids 1 and 2
w = the weight% of liquid 1 and 2

Note that this approach gives units that are equivalent to density units (mass/volume). The apparently easier method of just multiplying densities by the weight% of each liquid and dividing by 100 will give incorrect answers. While the results yielded by the easier method are sometimes close to the correct values, the difference will often be significant. In summary, when converting from one composition base to another, first of all convert the units (mass, volume or mols) to the new unit of measurement and then determine the composition.

5.6 HEAT BALANCES

When considering heat balances around parts of a process, three of the concepts discussed earlier are valid. They are as follows:

- The concept of "heat in must equal heat out" can be substituted for the similar material balance concept that "mass in must equal mass out." However, it should also include the idea that the heat removed by a stream must be equal to the amount of heat gained by another stream. An example of this is a fractionation tower where heat is rejected to a water-cooled condenser. The heat removed by the condenser must be equal to the heat gained by the cooling water stream. This heat gained by the cooling water stream will be rejected to the atmosphere as the cooling water is routed to a cooling tower and cooled by contact with air.

- The need to have a steady state condition in order to achieve a satisfactory heat balance is similar to the need to have steady state conditions in order to achieve a satisfactory material balance. The potential for accumulation or deaccumulation of heat is also present. An example of this is an exothermic (heat generated) reaction that is occurring inside a process vessel with no means to remove the heat generated. The vessel and its contents will increase in temperature, creating an unsteady-state accumulation of heat.

- The concept of the encircling line is valid for heat balances as well as for material balances. If a stream such as that described in Figure 5-3 is cut by an encircling line, the line must be included in the heat balance. If, as in the case of the recycle stream in Figure 5-3, it is not cut by the encircling line, it should not be included in the heat balance.

A general equation that represents a heat balance around a process vessel where an exothermic reaction is taking place is as follows:

$$Q_{AC} = Q_I + Q_G - Q_O - Q_R \qquad (5\text{-}10)$$

where

Q_{AC} = heat being accumulated in the process vessel (at steady state, this will be equal to zero)

Q_I = heat content of the incoming fluids

Q_G = heat generated by the reaction

Q_O = heat content of the fluids leaving the vessel

Q_R = heat removed by the cooling system

When equation (5-10) is applied, if the temperature in the process vessel is constant, there will be no heat accumulation in the vessel and Q_{AC} will be equal to zero. In that case, equation (5-10) can be rearranged as shown in equation (5-11) below:

$$Q_R = Q_I + Q_G - Q_O \qquad (5\text{-}11)$$

Equation (5-11) indicates that at steady state (the process vessel is at a constant temperature), the amount of heat that must be removed by the cooling system is equal to the amount of heat added by the incoming fluids plus the heat generated by the reaction, less the heat removed by the fluids leaving the vessel. Equation (5-11) can be used for two different types of calculations. It can be used to determine how much heat must be removed by the cooling system, and it can be used to determine how much reaction is occurring, if the heat being removed by the cooling system is known.

Since a high percentage of the reactions in the process industry are exothermic, the focus of this book is on this type of reaction. However, a very similar equation to equation (5-11) can be developed for endothermic reactions. These are reactions that require that heat be added in order to maintain the process vessel at a constant temperature.

While equations (5-10) and (5-11) appear to be conceptually relatively simple, there are obviously some questions regarding how the terms in the equations are developed. The following paragraphs are provided to describe how these terms are developed.

The heat content terms designated by the letter Q consist of a flow rate in mass/time units (e.g., lb/hr) and an enthalpy term in heat units per mass (e.g., BTU/lb). These two terms are multiplied together to give heat units/unit time (e.g., BTU/hr). Enthalpy is a thermodynamic term that is a measure of how much heat a unit mass (lb) of material contains. The English unit of enthalpy is the British thermal unit, or BTU. It is defined as the amount of energy required to raise the temperature of 1 lb of water 1 degree Fahrenheit. Enthalpy is a function of both specific heat and actual temperature. It can be found in handbooks, computer programs, or determined by calculations. The calculation technique helps to understand the meaning of the variable. It is usually calculated as follows:

$$H = C_{p} \times (t_{A} - t_{R}) \qquad (5\text{-}12)$$

where

H = enthalpy, BTU/lb
C_{p} = specific heat, BTU/lb-°F
t_{A} = actual temperature, °F
t_{R} = reference temperature, °F

Specific heat, in technical terms, is the amount of heat required to raise 1 lb of a material 1°F. It can be viewed as the sensitivity of the material to heat input. Water has a specific heat of approximately 1 BTU/lb-°F. Hydrocarbon liquids have a specific heat of 0.4 to 0.6 BTU/lb-°F. Nitrogen gas, on the other hand, has a very low specific heat, about 0.14 BTU/lb-°F. The reference temperature is an arbitrary value. It can be -200°F, 0°F or any other value. *The only key is that it must stay the same for any single heat balance.*

The enthalpy term that has been considered so far has only dealt with what is known as sensible heat. Sensible heat is the heat that is removed from or added to a system in which there is no change of state, that is, in which there is no vaporization, condensation, melting, or solidification. Examples that require only consideration of sensible heat include a home water heater, process cooler, or process heater. When there is a change of state, such as condensation of vapors or vaporization, there is a change in the enthalpy, which is referred to as latent heat. Thus condensation of a vapor such as propane might occur with no temperature change, but with a significant change in enthalpy, as the vapor changes state to a liquid. Similarly, the heat of

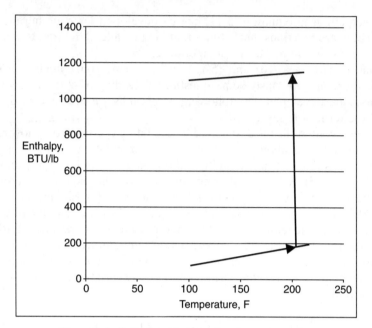

Figure 5-4 Enthalpy path to heat and vaporize water.

reaction is the heat generated when a single pound of material changes composition, and possibly state, from one component and/or state to another. Again, there may be no temperature change, but there will be a significant change in enthalpy.

These concepts are illustrated in Figure 5-4. In this figure, the enthalpy of water or steam is shown on the vertical axis and temperature is shown on the horizontal axis. The lower curve represents the enthalpy of water at its boiling point. The upper curve represents the enthalpy of steam at the boiling point of water. The actual pressure will also be a function of temperature. For example, at 212°F, we know that the pressure is atmospheric, or, 14.7 psia. At this temperature, the enthalpy of water is 180 BTU/lb and the enthalpy of the vapor (steam) is 1150 BTU/lb. By taking their difference, the latent heat of vaporization can be calculated to be 970 BTU/lb. Thus it will take 970 BTU to vaporize 1 lb of water if the water is at its boiling point. If the water is only at 100°F, sensible heat must be added to raise the temperature to 212°F. The amount of sensible heat required can be calculated using equation (5-12). In using equation (5-12), the enthalpy must be determined at the two temperatures (100° and 212°). Assuming a specific heat of water is 1 BTU/lb-°F and a reference temperature of 0°F, equations (5-13) and (5-14) show these calculations.

$$H_{100} = 1 \times (100\text{-}0) = 100 \text{BTU/lb-}°\text{F} \qquad (5\text{-}13)$$

$$H_{212} = 1 \times (212\text{-}0) = 212 \text{BTU/lb-}°\text{F} \qquad (5\text{-}14)$$

Thus the sensible heat required to heat the water to the boiling point is 112 BTU/lb and the total heat required to heat and vaporize a pound of water is 1082 BTU/lb (970+112). The overall path of this process is shown as a heavy line in Figure 5-4.

The purpose of this book is not to provide a source of all of the possible values of specific heat, latent heat of vaporization, or heat of reaction, but to illustrate how these values are used once they are obtained from other sources. It is likely that the procedure for determining the appropriate heat content of a process is an area in which a process operator may require assistance from a process engineer or a specialist in data resources.

EXAMPLE PROBLEM 5-2

The production process of polypropylene provides excellent examples of heat removal. The reaction is highly exothermic, which requires that heat be removed to ensure that the reactors are kept at a constant temperature. The feed streams are fed to the reactors at a temperature lower than the reaction temperature. Therefore, some of the heat of reaction is removed by the process of heating the feed streams to the temperature of the reactors. The reactor heat, though normally rejected to the atmosphere in a cooling tower, is rejected to cooling water. In this example, the following are given:

- Bulk polymerization (the only feed is propylene and a catalyst). The reactor is operated at a high enough pressure that propylene is in its liquid phase.
- Propylene is fed to the reactor at 100°F and a feed rate of 40,000 lb/hr.
- The specific heat of propylene is 0.60 BTU/lb-°F.
- The specific heat for polypropylene is 0.45 BTU/lb-°F.
- The heat of polymerization is 970 BTU/lb.
- The polymerization rate is 20,000 lb/hr, at a reactor temperature of 160°F.
- Heat is rejected to cooling water, which enters the exchangers at 90°F and exits the exchangers at 110°F.

Determine the total heat of polymerization, the amount of heat rejected to the cooling water, and the cooling water rate.

Step 1: Determine the total heat of polymerization. This can be done as shown in equation (5-15)

$$Q_G = 20,000 \times 970 = 19,400,000 \, \text{BTU/hr} \tag{5-15}$$

Step 2: Determine the heat rejected to the cooling water. This can be done using equation (5-11) and equation (5-12), and assuming a reference temperature of 0°F, as follows:

$$Q_R = Q_I + Q_G - Q_O \tag{5-11}$$

$$Q_I = (100\text{-}0) \times 0.6 \times 40,000 = 2,400,000 \tag{5-16}$$

$$Q_O = (160\text{-}0) \times 0.6 \times 20,000 + (160\text{-}0) \times 0.45 \times 20,000$$
$$= 3,360,000 \tag{5-17}$$

$$Q_R = 2,400,000 + 19,400,000\text{-}3,360,000$$
$$= 18,440,000 \text{ BTU/hr} \tag{5-18}$$

Step 3: Determine the cooling water rate. Knowing that the heat rejected to the cooling water is 18,440,000 BTU/hr, and that the cooling water with a specific heat of 1 BTU/lb-°F enters the exchanger at 90°F and leaves at 110°F, the cooling water rate can be calculated as follows:

$$WC = 18,440,000/((110\text{-}90) \times 1) = 922,000 \text{ lb/hr} \tag{5-19}$$

Since cooling water rates are traditionally expressed in gallons/minute, this value can be determined as follows:

$$WC_{gpm} = 922,000/(60 \times 8.34) = 1840 \text{ gpm} \tag{5-20}$$

Eventually, all of this 18,440,000 BTU/hr that is rejected to the cooling water must be rejected to the atmosphere, either by heating air or by vaporizing the water in the cooling tower that is used to cool the return water to 90°F.

5.7   FLUID FLOW

Any discussion of chemical engineering principles must cover fluid flow. The analyses of pump and compressor problems, as well as that of pipeline networks, all involve consideration of fluid flow. Bernoulli's equation is one of the keys to understanding fluid flow. It is shown below:

$$\Delta P/\rho + \Delta(v^2)/2g_c + \Delta z = -w - lw \tag{5-21}$$

where, in English units,

ΔP = pressure difference between two points, lb/ft^2

ρ = fluid density, lb/ft^3

$\Delta(v^2)$ = difference in velocity squared between two points, (ft/sec)^2

g_c = gravitational constant, ft/sec^2

Δz = difference in elevation between two points, ft

w = amount of work added by the prime mover, ft

lw = frictional loss in the piping system, ft

There are several key points to be learned from an inspection of this equation.

- The units are "feet of head." For example, when considering the units on the first term, lb/ft^2 divided by lb/ft^3 gives the units of feet. The use of feet units provides an easy means to include changes in elevation and velocity.
- For incompressible fluids (liquids) flowing in a pipe of uniform diameter with no work added and no elevation change:

$$\Delta P/\rho = -lw \qquad (5\text{-}22)$$

$$\text{or } \Delta P = -lw \times \rho \qquad (5\text{-}23)$$

Thus, equation (5-23) allows expression of the pressure drop in more conventional units of lb/ft^2, or, when divided by $144\,in^2/ft^2$, values of psi are obtained.

- Under certain conditions, the change in elevation will equal the frictional loss in the pipe, there will be no change in pressure head, and the pressure drop will appear to be zero.
- If there is a change in pipe diameter, there will be a direct change in pressure. That is, if the pipe diameter is increased at constant elevation and frictional loss can be ignored, the pressure will increase, since the velocity decreases.
- While Bernoulli's equation is valid for compressible fluids (gases), the change in density with pressure makes the analysis much more complicated.

The process designer of a new plant will optimize the piping design to allow for a balance of investment and operating cost. The larger the diameter of the pipe that is used, the lower the energy loss and, hence, the lower the operating cost, but the higher the investment. In addition to this investment/operating cost optimization, there may be other factors such as vacuum operation that dictate the size of the pipe. Typical flow velocities in a pipe are 4 to 6 ft/sec for liquids and 60 to 100 ft/sec for gases.

The industrial problem solver is often faced with a situation in which there appears to be an excessive pressure drop in a pipe. Thus, he needs to know if the calculated pressure drop is really lower than the observed pressure drop. The next area to be covered in this section deals with the calculation of frictional losses. Once the calculated pressure drop is known, it can be compared with the actual pressure drop. If the actual pressure drop is higher than the calculated pressure drop, there is a chance that there is a restriction in the line. All of the terms in Bernoulli's theory, equation (5-21) must be taken into account in the calculation of pressure drop. This is illustrated later in Example Problem 5-3.

Liquids in a pipe can have either turbulent or streamline flow; these names are descriptive of the type of flow. Streamline flow is smooth, with a well-defined flow pattern across the diameter of the pipe. The appearance is much like the flow appearance of the deep part of a river. On the other hand, turbulent flow consists of an undefined pattern, much like that of the rapids in a river. In the process industry, viscosity and flow velocity are the biggest factors in determining whether a liquid is in turbulent flow or streamline flow. Viscosity is a fluid property that measures how flowable a liquid is. For example, lubricating oil has a relatively high viscosity and flows very slowly, compared with gasoline or water. With the exception of high-viscosity liquids, most industrial flow is in the turbulent category. When considering a liquid flowing in a pipe with turbulent flow, frictional pressure losses can be calculated using equation (5-24). While this equation does not appear to be dimensionally consistent, the dimension conversion factors have been included in the constant of 0.323:

$$\Delta P = 0.323 \times f \times S \times U^2 \times L/D \qquad (5\text{-}24)$$

where

ΔP = pressure drop, psi

f = a friction factor that can be approximated from the equation below:

$$f = -0.001 \times \ln(D \times U \times S/Z) + 0.0086 \qquad (5\text{-}25)$$

S = specific gravity relative to water (can be obtained either from handbooks or by dividing the density of the fluid by the density of water)

U = velocity in the pipe, ft/sec (can be calculated by dividing the flow rate in ft^3/sec by the cross sectional area of the pipe, ft^2)

L = pipe equivalent length, ft

D = pipe diameter, in

Z = fluid viscosity, centipoises

$\ln(D \times U \times S/Z)$ = natural logarithm of the term $(D \times U \times S/Z)$

The "equivalent pipe length" is a term that takes into account fittings such as tees and elbows. These short runs of pipe cause much more pressure drop than an equivalent length of straight pipe. An approximation of the equivalent length can be made by doubling the actual linear length.

While equation (5-24) was developed for liquids which are not compressible, it can be used for compressible gases provided that the pressure drop is less than 10% of the initial pressure. If the pressure drop is greater than this, more complicated techniques to determine the pressure drop in gases will be required. This is an area in which assistance of a graduate engineer will likely be necessary.

In addition to frictional losses in a pipe with uniform cross-sectional area, there are also frictional losses when a process fluid enters or leaves a process vessel. These losses are known as expansion or contraction losses, respectively. They are usually small, relative to the losses encountered in the pipe, although occasionally they do need to be considered. Assuming that the ratio of the pipe area to the vessel area approaches zero, the pressure drop due to expansion (flow entering the vessel) and pressure drop due to contraction (flow leaving the vessel) can be determined by equations (5-26) and (5-27), respectively.

$$\Delta P_C = 0.5 \times U^2 \times S/148.2 \tag{5-26}$$

$$\Delta P_E = U^2 \times S/148.2 \tag{5-27}$$

where

ΔP_C = pressure drop due to friction in contraction, psi
ΔP_E = pressure drop due to friction in expansion, psi
U = velocity in the pipe, ft/sec

These relationships can be used for either liquid or gas flow. The relationships can also be used to calculate the expansion or contraction loss when the pipe diameter changes. They will give elevated values, due to the assumption of the pipe-to-vessel diameter ratio being zero. Generally, the expansion and contraction loss when the pipe diameter is changed can be ignored, except for the case of exceptionally high flow velocities.

The relationships described above are valid as long as the calculated gas velocity does not exceed a value that is referred to as either sonic or acoustic velocity. Sonic velocity has been defined as the velocity of sound in a gas. For example, the velocity of sound in the atmosphere is about 1100 ft/sec. Thus the distance between a location and a lightning strike can be gauged by counting the number of seconds between the observation of the flash and the sound of the thunder and multiplying that time by 1100. In fluid flow, sonic velocity represents the maximum velocity a gas can obtain in a pipe of constant diameter or across an orifice. With a series of converging and diverging nozzles, a velocity greater than the speed of sound can be obtained. This design principle of converging and diverging nozzles is used in steam jets to achieve supersonic velocity. However, it is beyond the scope of this book to cover this area. What is important is to recognize that there is a maximum flow that can be achieved in a pipe or across an orifice. While there are calculation procedures for determining the actual sonic velocity in any fluid, most applications of this concept use the calculated ratio of the downstream and upstream pressures to determine if sonic velocity is being approached. A general rule is that sonic velocity is occurring across an orifice if the ratio of the downstream pressure to upstream pressure is less than 0.52. This is shown mathematically in equation (5-28) below:

$$P_D/P_U > 0.52 \tag{5-28}$$

where

P_D = pressure downstream of the orifice

P_U = pressure immediately upstream of the orifice

If the ratio is less than 0.52, the flow and pressure drop calculation procedures discussed earlier are not applicable.

The most common area in which this concept is applied is in the design of the orifice in safety valves. The orifice is designed to release whatever quantity is necessary to avoid over-pressuring the vessel. Essentially, all safety valves discharge to a low pressure (P_D). The upstream pressure immediately ahead of the orifice (P_U) that is in the safety valve is often much greater by several hundred psia. For example, if the discharge pressure of a safety valve is 20 psia and the upstream pressure is 150 psia, the ratio is as shown in equation (5-29) below:

$$P_D/P_U = 20/150 = 0.13 \tag{5-29}$$

Therefore, since the ratio is below 0.52, sonic velocity will occur across the orifice and a different calculation approach must be used in sizing the orifice. This is also an area in which the assistance of a graduate engineer or someone trained in sonic flow calculations will be required.

EXAMPLE PROBLEM 5-3

Water is flowing through a pipe with diameter of 3 in and length of 200 ft, at a rate of 40,000 lb/hr, at 70°F. What is the total pressure drop in the pipe? What is the "exit pressure drop" as the liquid leaves the storage vessel and enters the pipe? Assume that water has a viscosity of 1 centipoise and a density of 62.4 lb/ft^3.

Step 1: Calculate the flow velocity in the pipe:

$$\text{Volumetric flow} = 40,000/(62.4 \times 3600) = 0.178 \text{ ft}^3/\text{sec} \tag{5-30}$$

$$\text{Area of pipe} = \pi \times 3^2/(4 \times 144) = 0.0491 \text{ ft}^2 \tag{5-31}$$

$$\text{Flow velocity} = U = 0.178/0.0491 = 3.6 \text{ ft/sec} \tag{5-32}$$

Step 2: Calculate the friction factor from equation (5-25):

$$DUS/Z = 3 \times 3.6 \times 1/1 = 10.8 \tag{5-33}$$

Since the density of water was given as 62.4, the specific gravity (S) of water is equal to 1.

$$f = -0.001 \times \ln(D \times U \times S/Z) + .0086 \qquad (5\text{-}25)$$

$$f = -0.001 \times \ln(10.8) + 0.0086 = 0.00622 \qquad (5\text{-}34)$$

Step 3: Calculate the pressure drop in the pipe using equation (5-24):

$$\Delta P = 0.323 \times f \times S \times U^2 \times L/D \qquad (5\text{-}24)$$

$$\Delta P = 0.323 \times 0.00622 \times 1 \times 3.6^2 \times 200 \times 2/3 = 3.47 \text{ psi} \qquad (5\text{-}35)$$

Note that the factor of 2 is included, to take into account the nonstraight run fittings (valves, elbows and tees).

Step 4: Calculate the pressure drop in the exit of the storage vessel using equation (5-26):

$$\Delta P_C = 0.5 \times U^2 \times S/148.2 \qquad (5\text{-}26)$$

$$\Delta P_C = 0.5 \times 3.6^2 \times 1/148.2 = 0.043 \text{ psi} \qquad (5\text{-}36)$$

This would be a fairly typical result. That is, the exit loss is generally small compared to the overall pressure drop. However, there are times when this exit loss is significant in problem solving.

The steps shown above give approximate answers which are generally sufficient for problem solving. However, the inherent errors in the procedures given are as follows:

- The empirical approach of doubling the actual length of pipe to take into account fittings is not always correct. If an accurate calculation is required, it will be necessary to count the number of each type of fitting and use handbook values to obtain the equivalent length of each fitting, and, hence, the exact total equivalent length of the pipe.
- A nominal 3-in pipe has an inside diameter that can be greater or less than 3 in, depending on the type and thickness of the pipe material. There are handbooks that provide the actual inside diameter of various types and pressure ratings of pipe.
- The equation for calculating the friction factor gives an approximate factor for commercial pipe. More accurate tables are available in handbooks that give the factor as a function of pipe roughness.

5.8 EQUILIBRIUM AND RATE

Operations other than fluid flow in the process industry can either be "equilibrium limited or "rate limited." This is true whether the process being considered is a chemical reaction or a physical operation such as heat exchange, fractionation, adsorption, or drying. It is important to be able to understand and differentiate between the two limitations.

When attempting to understand equilibrium, an example from the day to day world might help. Think about the following questions:

- Is the earth in equilibrium with the sun?
- Is the earth getting continuously closer to or farther away from the sun? Think in terms of distances that can be easily measured on a day to day basis.

If it is believed that the earth is approximately the same distance from the sun as it was at this time last year, then it makes sense to say that the earth and sun are in equilibrium. That is, the forces that would tend to push the earth farther away from the sun or pull it closer to the sun are in balance.

Similar concepts can be applied to the process industry. In the process industry, equilibrium is a state in which there is no further change in properties, regardless of the amount of time provided. Examples of equilibrium from the process industry are:

- *Concentration of a solvent in the vapor space*: In a nitrogen-blanketed tank containing a solvent, the concentration of solvent in the vapor space will not be higher than an equilibrium amount that is a function of the temperature of the solvent.
- *Equilibrium-limited reactions*: Many chemical reactions are equilibrium-limited, meaning that the conversion of one component to another will only go to a certain level regardless of how long the reaction is allowed to continue. Again, this equilibrium can be shifted by temperature and/ or the initial concentration of one of the reactants.
- *Removal of solvent from a polymer*: The drying or removal of a solvent from a polymer will sometimes be limited by the equilibrium between the solvent phase in the polymer and the solvent in the vapor. Again, this can be shifted by changes in temperature. It should be recognized that it is rare that an equilibrium concentration can be zero regardless of the temperature.
- *Dissolving of a material in a solvent*: The saturation concentration of a solute in a solvent is the equilibrium concentration at that temperature. A real world example of this principle occurs when sugar is dissolved in tea or coffee. Regardless of how hard one stirs, there is a limit to how much sugar can be dissolved.

The previous comments give a brief summary of the concept of equilibrium. Many of these concepts are discussed in more detail in subsequent chapters.

One area which requires further explanation before the information in the subsequent chapters is encountered is the concept of vapor-liquid equilibrium. The understanding of vapor-liquid equilibrium is fundamental to many key concepts, such as the understanding of the estimation of the concentration of a solvent in the atmosphere and the understanding of the removal of a solvent from a polymer, as well as fractionation and other operations during which there is an interface between liquid and vapor phases or between polymer and vapor phases. There is some terminology that must be learned to fully understand this form of equilibrium. The key terms are as follows:

- Vapor pressure is one of the key factors in determining how much of a liquid will be in the vapor phase. For example, at the boiling point of a liquid at atmospheric pressure, the vapor pressure is equal to 14.7 psia, or 0 psig. At its boiling point, a liquid will completely vaporize, given sufficient time and heat input. Below the boiling point, the higher the vapor pressure, the higher will be the concentration of liquid in the vapor phase. Water at 100°F has a vapor pressure of 0.949 psia. Thus the mol fraction of water in the vapor immediately above a tank of water at 100°F is 0.0646 (0.949/14.7). Even though the water is more than 100°F below the boiling point, there will be water vapor in the vapor phase. Vapor pressure values can be obtained from handbooks or computer programs.

- Partial pressure is proportional to the amount of a compound in the vapor phase. It is also related to the vapor pressure and the amount of the compound in the liquid phase. Mathematically it can be expressed as follows:

$$PP = Y \times \pi = X \times VP \qquad (5\text{-}37)$$

where

PP = partial pressure, psia

π = total pressure, psia

VP = vapor pressure, psia

Y = vapor phase concentration, mol fraction

X = liquid phase concentration, mol fraction

- The equilibrium constant (k) is the ratio of the vapor phase composition to the liquid phase composition, using mol fractions. This can be expressed mathematically as follows:

$$k = Y/X \qquad (5\text{-}38)$$

The value of the constant (k) can often be determined from tables, charts, or computer programs. If the value is not available, it can be approximated as shown in equation (5-39) below:

$$k = VP/\pi \qquad (5\text{-}39)$$

Once k is determined from tables, charts, computer programs, or from equation (5-39), equation (5-38) can be used to obtain a relationship between the vapor and liquid phase composition. Similar relationships can be developed for polymer-vapor equilibrium; these relationships will be more complex. They are discussed in detail in Chapter 13.

The discussion of equilibrium between phases discussed above is based on ideal systems. The approach works reasonably well for a pure component or for mixtures that contain only one type of component—hydrocarbons or alcohols, for example. When considering a mixture of different types of compounds there will almost always be some "nonideality." For example, at a fixed temperature, the k value for small amounts of methanol dissolved in hexane will almost always be higher than the k value for methanol available from charts, tables, computer programs, or calculated using equation (5-39). These mixtures are referred to as "nonideal solutions." Water with almost any other component will form a nonideal solution.

Once the equilibrium concentration in a process is known, there will always be the question of whether this concentration be achieved. The speed at which a solution moves from a given concentration to the equilibrium concentration is known as rate. It can be rate of reaction, rate of drying, or rate of mass transfer. As described in Chapter 9, this rate will be a function of a constant that incorporates surface area, mass transfer rates and driving force. The driving force represents the difference between the equilibrium value and actual value, and is often dictated by the process conditions.

On the other hand, the constant can be strongly influenced by equipment design. For example, if a design calls for cooling a process liquid from 120°F to 95°F, using cooling water at 90°F, which then leaves the heat exchange equipment at 110°F, the driving force is fixed. The difference in temperatures through the exchanger is the driving force, and inlet and outlet conditions for both the hot and cold sides are specified by the design. However, the surface area and the physical design of the exchanger can be varied to improve the speed of heat transfer. Therefore, the speed of heat transfer is influenced by the type of equipment to be utilized.

Table 5-1 shows the relationship between some selected variables and their effect on equilibrium, rate, or both. For example, temperature and pressure can influence both equilibrium and rate. However, catalyst and equipment can only influence rate. While rate-limited processes are dealt with extensively in chapters 8 and 9, some examples of how equipment design can influence rates are described in the following paragraphs.

Table 5-1   Variable influences

Variable	Equilibrium	Rate
Temperature	x	x
Pressure	x	x
Catalyst		x
Equipment design		x

Fractionation columns are analyzed with the assumption that there are multiple equilibrium stages in each column. The term "tray efficiency" is used to gauge how close the composition of each tray approaches to its equilibrium. A tower with a tray efficiency of only 50% would require two actual trays to reach equilibrium between the vapor phase and liquid phase. Thus if the tray could be designed to provide a tray efficiency of 100%, the number of actual trays could be reduced by a factor of 2. While it is unlikely that 100% tray efficiency can be achieved, improved design of both the vapor-liquid contacting region of the tray and the vapor-liquid separation region of the tray can improve tray efficiency. This is an example of how fractionation equipment design can impact the rate at which equilibrium is achieved.

The rate at which heat is transferred into a cooler can be gauged by the approach of the hotter fluid-outlet side temperature to the cooler fluid-inlet side temperature. This is because the only equilibrium limit in heat exchange is that the hot side outlet temperature cannot be cooler than the cold side inlet temperature. Plate and frame heat exchangers, by nature of their mechanical design, allow for a very close approach compared to that of more conventional shell and tube heat exchangers.

In certain rate-limited reactions, the speed of the reaction can be enhanced by changing the size of the catalyst particle in fixed-bed reactors. As the catalyst particle size is decreased, the catalyst particle surface area increases. As more surface area is available, there are more catalyst sites accessible to the reactants.

These are three examples of reaction rate being enhanced by equipment design without any changes in temperature or pressure.

EXAMPLE PROBLEM 5-4

Assume that hexane liquid at 85°F is in a vessel with a nitrogen atmosphere in the vapor phase. The total pressure on the drum is 5 psig. Determine the equilibrium constant value of hexane at 85°F, and the concentration of hexane in the vapor phase. Vapor pressure data indicates that the vapor pressure of hexane at 85°F is 3.7 psia.

Step 1: Determine the equilibrium value using equation (5-39).

$$k = VP/\pi \qquad (5\text{-}39)$$

$$k = 3.7/(5+14.7) = 0.19 \qquad (5\text{-}40)$$

Step 2: Use the calculated equilibrium constant (k) to calculate the vapor phase composition with the modified equation (5-38) below. This approach assumes that there is essentially no nitrogen dissolved in the hexane. Thus the liquid phase has a mol fraction of hexane of 1.

$$Y = k \times X \tag{5-41}$$

$$Y = 0.19 \times 1.0 = 0.19 \text{ mol fraction of hexane in the vapor} \tag{5-42}$$

As indicated in this section, a process does not always achieve equilibrium. However, when considering vapor-liquid equilibrium of low viscosity liquids, it is almost always correct to assume that equilibrium is reached in a process vessel with good vapor-liquid disengaging characteristics.

6

DEVELOPMENT OF WORKING HYPOTHESES

6.1 INTRODUCTION

The title of this chapter requires some explanation. Dictionaries define a hypothesis as "a theory needing investigation" or "a tentative explanation for a phenomenon used as a basis for further investigation." A working hypothesis is just that: It is a tentative explanation that can be used to investigate a problem further. This book does not deal with the multitude of methods that can be used to generate potential working hypotheses. Many of these methods put almost complete emphasis on accurate problem statements. The implicit assumption is that if the problem can be defined in sufficient detail then the problem solution will be apparent. In complex process plants, multifaceted problems can rarely be solved through this simple approach.

What is presented in this chapter is an approach for the development of theoretically sound working hypotheses based on careful consideration of a series of questions. These questions will require an analysis of the data in an introspective fashion. It is highly unlikely that the approach described in this chapter will provide only one possible problem solution. Thus, this and subsequent chapters also deal with ways in which the large number of possible working hypotheses can be narrowed down using logic and, most importantly, one's technology training.

Problem Solving for Process Operators and Specialists, First Edition. Joseph M. Bonem.
© 2011 John Wiley & Sons, Inc. Published 2011 by John Wiley & Sons Inc.

6.2   AREAS OF TECHNOLOGY

As indicated earlier, there are two types of technologies utilized in process plants. These are process-related technologies and equipment-related technologies.

Process technologies are technologies that deal with a specific process. Examples of these include a polymerization process, an isomerization reactor process, and a distillation process. The process-specific technologies may include items such as reaction rate kinetics, polymer product attributes, or relative volatility data. Each of these processes will have specific technology details which must be well known and understood by the problem solver before he attempts to do any process-technology-related problem solving.

Equipment-related technologies are technologies that are valid for any specific piece of equipment, regardless of the process technology in the facility in which it is being used. Examples of these are details and calculations associated with pumps, compressors, and distillation towers. In addition, most kinetically limited processes (e.g., heat transfer) can be generalized in terms which will allow a hypothesis to be developed regardless of the specific technology. This approach to kinetically limited processes is described in detail in Chapter 9.

6.3   FORMULATING HYPOTHESES VIA KEY QUESTIONS

The primary purpose of Chapters 7 through 10 is to demonstrate how to use the five-step problem-solving procedure discussed in Chapter 3. The other equally important parts of this procedure, having a daily monitoring system and determining the optimum technical depth, were adequately covered in Chapters 3 and 4.

The emphasis in this chapter is on formulating and verifying theoretically sound working hypotheses. Formulating theoretically sound working hypotheses deals with an in-depth thought process that requires the problem solver to utilize his engineering training to develop a hypothesis. The in-depth thought process is rarely done in meetings. It often requires data analysis, literature research, and/or "one on one" discussions with experts in the field or those who can serve as a source of data. This in-depth thought process often involves consideration of the potential questions shown below to help define the cause of the problem. Examples are given for each question. These examples are not meant to be an inclusive list, but are given only to amplify the specific question. The questions are given in order of priority. Obviously this priority must depend on the specific problem and the specific process. These questions also assume that steps 1 (verify that the problem actually occurred) and 2 (write out an *accurate* statement of what problem you are trying to solve) have been completed.

1. Are all operating directives and procedures being followed? An inspection of operating conditions at most process plants will show that deviations from procedures and directives are occurring. These may or may not be related to the problem of interest. While it is important to verify that all procedures and directives are being followed, a small deviation should not be deemed to be the source of the problem unless there is a theoretically sound working hypothesis which explains how the deviation is causing the problem.

2. Are all instruments correct? Incorrect flow meters may result in reaction rates being different than expected, fractionation separation being below design levels, a pump or compressor appearing to be operating "off the curve," or failure to adequately strip an impurity from a polymer. A level instrument being wrong could result in reduced or increased reaction rate, which would manifest itself as a lesser or greater amount of reaction. Heat and material balances are exceptionally good tools and can often be used to answer this question.

3. Are laboratory results correct? If the problem under study is related in any way to laboratory results, confirming that the laboratory results are correct is a high priority. This confirmation can require review of the procedures as written, review of the procedures as performed, and review of the chemicals being used. For example, the results of an extraction procedure can be greatly altered if cyclohexane is used as the solvent when the procedure calles for the use of normal hexane.

4. Were any errors made in the original design? The high priority given to this possibility is due to the need to assess this question early in the problem-solving activity, prior to doing a large amount of work in other areas. This assessment can be made based on experience with the process and the length of time it has been in operation. The probability that original design errors are causing the operating problem decreases with the age of the process. However, one should not assume that just because a process has been in operation for several months that it is free of design errors. A new operating condition or new product grade may expose design errors that were not detected earlier. These design errors might be as small as a density being wrong on an instrument specification sheet to as large as incorrect tray selection for a distillation column. The assessment of potential design errors can only be made by either a detailed review of design calculations or by redoing these calculations. If it is necessary to review or redo design calculations, the process operator may need to obtain assistance from an expert skilled in this area.

5. Were there changes in operating conditions that occurred at the same time the problem began? These changes in operating conditions may immediately result in the observation of a process problem. However, the most likely scenario is that everything appears normal when the

changes are first made. However, at some later point in time there will be a small change in another variable and the problem becomes noticeable. An example of this might be a reduction in the operating temperature of an exothermic reactor in winter. After the reduction in temperature, all control systems appear to be operating normally. However, temperature control is impossible as the outside temperature and, hence, cooling water temperature increases with spring conditions. If the process being investigated is integrated with other processes (e.g., in a refinery or chemical plant complex), it will be desirable to also investigate operating condition changes in these other processes.

6. Is fluid leakage occurring? The term "fluid leakage" covers such areas as leakage through heat exchanger tubes, leakage across isolation valves, and leakage through control valves. Leakage can cause reaction rates to be lower than desired if the leaking component is an impurity. Leakage can also cause an apparent loss of fractionation efficiency or an apparent loss of pumping or compression efficiency. The potential for leakage can be determined by a flow sheet review, determination of pressure flow potential using measured pressures, and, in some cases, detailed calculations of control valve clearances.

7. Has there been either normal or unusual mechanical wear or changes that could impact performance? Erosion of wear rings or failure of check valves can greatly affect the performance of pumps and compressors. The performance of distillation columns can suffer due to the failure of a single tray segment. Unusual mechanical wear will often be caused by a large process upset. Mechanical changes might occur that would affect the process performance. For example, a change in the composition of material used in a mechanical seal might result in decomposition of the seal and contamination of a product.

8. Is the reaction rate as anticipated? At times, undesirable reaction rate is the problem . However, there are also times when an excessive amount of reaction or a lack of reaction is the cause of the problem. For example, alumina desiccant is known to have catalytic properties. The catalytic properties can usually be mitigated by operating techniques. However, if a batch of alumina desiccant has exceptionally high catalytic activity, the normal compensatory operating technique may not be adequate. This may cause problems in the process that can be traced back to the specific batch of alumina.

9. Are there any adverse reactions occurring? Adverse reactions are always a potential problem when dealing with reactive chemicals. The presence of solids in a distillation column that purified a diolefin, such as butadiene or isoprene, might be traced to small quantities of oxygen that entered the process from inadequately purged storage vessels and then catalyzed the reaction of the diolefin to form a polymer. Another example is the utilization of aluminum metal in an analyzer sample

system in which chlorides and olefins are present. The chlorides would react with the aluminum to form aluminum chloride, which would then cause the olefin to polymerize to an oily material and foul the analyzer. The trays in a distillation column might become plugged due to solids formed by the unexpected presence of water.

10. Were there errors made in the construction of the process? Construction errors are treated as a low priority simply because they almost always involve a unit shutdown or elaborate, noninvasive techniques to inspect potential problems. However, these errors do occur. Examples of construction errors are debris left in vessels or piping, improper leveling of distillation trays, beveled orifice meters installed backwards, failure to complete the cutting of a "hot tap" so that the flow path is greatly restricted, and installation of an incorrect pump impeller.

This list is meant to serve as a possible checklist that can be used to develop hypotheses. It is not meant to be an all-inclusive list. Certainly the priorities will change depending on the specific process and status of the process. However, it is believed that this approach can be effective in developing sound hypotheses.

While the crux of this book is directed toward equipment-related technologies, this approach to developing working hypotheses can be applied to process technologies as well.

6.4 BEAUTY OF A SIMPLIFIED APPROACH

The problem solver will often be tempted to invent a very complicated theory to explain observations that he does not fully understand. Generally, this will be counterproductive. A simple theory will often lead to a problem solution that is easier to execute and more effective. Over 100 years ago, the physicist Ernest Rutherford commented, "A theory that you can't explain to a bartender is probably no good."

A real-life example of this concept is found in the history of the Panama Canal. Yellow fever and malaria were serious problems which claimed many lives, especially during the French period of canal building. Since the worst epidemics seemed to start during the rainy season, the initial theory was that these diseases were caused by mysterious vapors which formed and came out of the swamps during the rainy season. This complex theory provided essentially no problem solution. Prior to the American construction of the canal, Dr. Walter Reed and his coworkers in Cuba had developed a simple competing theory to the "swamp theory." They theorized that yellow fever was spread by the Stegomyia mosquito. In addition, they discovered that this mosquito would only lay eggs in clean water held in an artificial container located in or near a building occupied by humans. With this relatively simple theory, problem solutions became apparent. This allowed the yellow fever threat in Panama to be greatly mitigated and was one of the keys for successful completion of the Panama Canal. Similar approaches were developed to mitigate the malaria threat.

6.5 VERIFICATION OF PROPOSED HYPOTHESES

In addition to the technique described above, there are multiple alternative problem-solving techniques touted throughout the industrial world. All of them involve developing theoretically sound working hypotheses. These alternative techniques also put emphasis on maximizing the number of possible hypotheses, based on the desire to not overlook any possibility. The techniques are of value for generating possible hypotheses; unfortunately, they often result in the assumption that a problem solution that seems logical is also technically correct. This is not always so. The process by which to verify a proposed working hypothesis is a procedure wherein the hypothesis is carefully examined, using the best available techniques. This verification process can rarely be done without calculations. For example, the hypothesis of an operator that the restriction to the flow of a liquid in a 4-in line is due to a short section of 2-in pipe may or may not be technically correct. It can only be assessed as a theoretically sound working hypothesis by calculations. The input of the operator is very valuable, since this detail might have been overlooked. However, his conclusion is likely erroneous.

Regardless of how hypotheses are developed, they will require verification. It is imperative that the problem solver use his training and calculations to eliminate the hypotheses which are not theoretically correct. These required calculations are part of step 3 (develop a theoretically sound working hypothesis that explains the problem) or step 4 (provide a mechanism to test the hypothesis). Regardless of whether they are included in step 3 or step 4, the calculations should be done before proceeding with a plan for a plant test or operating changes.

The problem solver is often limited in his problem-solving ability by the lack of proven industrial calculation techniques. For example, pragmatic utilization of pump curves is not taught in the academic world and may not be well understood by process operators and technicians. However, it is exceptionally important in industry. To aid the problem solver, Chapters 7–13 contain hints and useful industrial calculation techniques and approaches. These sections will not include of all the possible technology and knowledge associated with various types of equipment, but are meant to be a summary of valuable techniques and knowledge for the process operator or specialist with plant problem-solving responsibilities.

Rather than discussing specific unit operations, the approach of this book is to discuss formulation of working hypotheses in the following four areas:

- Prime Movers (pumps and compressors)
- Staged Processes (towers)

- Kinetically Limited Processes (heat transfer, reaction, drying)
- Unsteady State Processes

It is recognized that most operating companies have technical manuals that describe how to design particular pieces of equipment. These manuals are usually voluminous and are aimed at design as opposed to problem solving. As such, the problem solver does not often use them. In Chapters 7–10, an attempt is made to reduce the important equipment concepts to those that are required to solve problems. The judgment as to what should be included in these concepts is obviously informed by the author's own experience. An attempt has been made to cover a broad spectrum of process equipment. If a specific type of equipment is not mentioned, there will be similar equipment described in sufficient detail to allow one to formulate a working hypothesis for the equipment.

6.6 ONE RIOT, ONE RANGER

The Texas Rangers are the oldest law enforcement group with statewide jurisdiction on the North American continent. According to Texas folklore, in the late 1800s there was a small Texas town in which a large riot was imminent. The town had limited law enforcement resources, so they requested that the governor of Texas send a troop of Texas Rangers to quell the riot. When the train arrived, carrying what the townspeople thought would be a troop of Rangers, only one person emerged from the train. When the townspeople expressed shock, the Ranger's reply was, "One riot, one ranger."

While it is true that a committee will always have more knowledge of facts than any single individual, the most effective problem solving is accomplished when one person has the responsibility to solve the problem and simply uses input from others. This single individual in process plants is almost always a chemical engineer, a process engineer, or an individual acting in the one of these roles. This individual can also be a process operator or specialist who has been assigned the responsibility to obtain a solution to a chronic problem. It is extremely important for him to bridge the chasm between his own training/ experience and that of others who may have important knowledge or data that will help solve problems. For example, very few process operators have detailed knowledge of mechanical seals. However, mechanical seals are an integral part of a centrifugal pump. Therefore, when a problem solver is working on a centrifugal pump problem, it is likely that knowledge of mechanical seals will be required. This knowledge can best be obtained from discussions with mechanics and/or mechanical engineers.

An example of the need to bridge the chasm between engineering disciplines was a chemical engineer who was asked the question "Is it okay to operate these two identical centrifugal pumps in parallel to obtain increased capacity?" His reply was "I don't know; that is an equipment question." In fact, the chemical engineer or process specialist will be the individual who has the necessary knowledge of process conditions to answer the question. He will have to conduct a more detailed analysis and develop a better understanding of centrifugal pumps to finalize the answer to the question. For example, he would have to develop understanding of the following.

- Running pumps in parallel involves determining the exact location on the pump curve (flow rate), determining the capability of the control system to handle the increased pressure drop, and determining the minimum flow that might be encountered during parallel pump operation. These are clearly more than "equipment questions."
- Running pumps in parallel at low flow rates can sometimes result in one of the pumps operating in a "blocked in" condition. If this condition occurs, one of the pumps is pumping the full process flow while the output from the other pump is very little. This "blocked in" operation could create a serious operation and safety problem. This condition could occur even though the pumps are identical, due to different clearances on the wear rings. This clearly requires consultation between various engineering disciplines.
- Although it is a little known fact, centrifugal pumps often experience a loss of stability similar to that which occurs in centrifugal compressors operating at lower flow than the surge point. If a centrifugal pump is operated at flows below this stability point, serious vibration can occur. The chemical engineer would certainly have to depend on another discipline to determine the stability limit.

The primary point of this discussion is that the process operator acting as a chemical engineer and being the "problem solver" in a process plant must know about or obtain knowledge of the equipment that is involved in the problem he is solving. He will often be required to determine why the piece of mechanical equipment does not have as much capacity as it should, rather than just saying "It is not operating at design capacity."

Similar points could be made in the areas of process chemistry. A process operator serving as a problem solver must obtain the necessary knowledge of the process chemistry in order to formulate theoretically sound, working hypotheses. Once the process chemistry is understood, the same five-step procedure can be used to solve process chemistry-related problems as well as process equipment problems. Time spent with a chemist familiar with the specific process technology chemistry will greatly aid in formulating correct working hypotheses. This need will be obvious in a process where reactors are utilized. However, in almost any process, the possibility for reaction exists. Ignoring the need to completely understand process chemistry can lead to the formulation of incorrect hypotheses. For example, in a polymer plant, a process engineer developed a hypothesis that the polymer was turning a light shade of pink due to the presence of iron complexes caused by corrosion. A more detailed analysis with a process chemist revealed that one of the additives being used was also a pH indicator, and that the pink color was due to the slight basicity of the polymer.

In spite of the emphasis on the "one riot, one ranger" approach above, the problem solver should never assume or pretend that he has all the knowledge necessary to work the problem. A truly effective problem solver will know and admit his limits. He will then look for help in areas outside of his knowledge. However, he will seek this knowledge in order to apply it to the current problem, as opposed to trying to reassign the problem to someone else.

7

APPLICATION TO
PRIME MOVERS

7.1 INTRODUCTION

Prime movers (pumps and compressors) are the workhorses of the process industry. They can be split into two generic categories, depending on how they impart energy to the process fluid. These categories are as follows:

- *Kinetic or Dynamic Systems*: Energy is imparted by accelerating the fluid with an impeller and then decelerating the fluid with an inherent rise in pressure. Examples of this type of equipment are centrifugal pumps and compressors, axial and centrifugal fans, and blowers. Bernoulli's theory can be used to understand how these prime movers operate.
- *Displacement Systems*: Energy is imparted by displacing the fluid from a lower pressure to a higher pressure environment with a pushing or rotating action. Examples of this type of equipment are reciprocating pumps and compressors and rotary blowers and compressors.

In order to formulate theoretically correct working hypotheses for problems with pumps and compressors, knowledge of several relationships and definitions are required. These are summarized in this chapter. In addition, several problems are presented to illustrate the use of these concepts.

Problem Solving for Process Operators and Specialists, First Edition. Joseph M. Bonem.
© 2011 John Wiley & Sons, Inc. Published 2011 by John Wiley & Sons Inc.

7.2 KINETIC SYSTEMS

As indicated above, Bernoulli's theory is the key to understanding these systems. The application of Bernoulli's theory to the flow of fluids in pipes was discussed in Chapter 5. In this chapter, it was only discussed in terms of the flow of fluids in piping. However, it is also valuable when considering the flow of fluids inside a pump or compressor case. Bernoulli's equation and the definition of terms is repeated as follows:

$$\Delta P/\rho + \Delta(v^2)/2g_c + \Delta z = -w - lw \qquad (7\text{-}1)$$

where

ΔP = pressure difference between two points

ρ = fluid density

$\Delta(v^2)$ = difference in velocity squared between two points

g_c = gravitational constant

Δz = difference in elevation between two points

w = amount of work added by the prime mover

lw = frictional loss in the piping system

When one is considering what happens inside the case of a centrifugal pump or compressor, only the first two terms of equation (7-1) are significant. They clearly indicate what happens to the fluid after it is accelerated by the work of the prime mover to approach the impeller tip velocity. As the fluid enters the outlet section of the pump case, it decelerates and the pressure increases to the discharge pressure. Figure 7-1 shows the typical flow path in a centrifugal prime mover.

Bernoulli's theory as applied to centrifugal prime movers has several implications that must be considered: The head ($\Delta P/\rho$) that a pump or compressor develops is directly related to impeller tip speed squared. High-speed impellers are required to develop high heads for either pumps or compressors. An alternative is multiple-staged impellers. The high-speed impellers and multiple-staged impellers can also be combined in a single pump or compressor.

The operating characteristic of a centrifugal system (pump or compressor) is defined by the equipment's characteristic "head curve." This head curve is developed by the equipment supplier and is provided as part of the equipment purchase. An example is shown in Figure 7-2. In this figure, head and horsepower are shown as a function of flow rate. The stability limit is also shown. Several generalized points can be made about this figure.

Operation to the left of the stability limit will result in flow instabilities, as flow surges forward and then backward through the prime mover. The stability is usually well defined for compressors and blowers. However, for pumps, it is usually 25–40% of the BEP (best efficiency point). If flow reversal

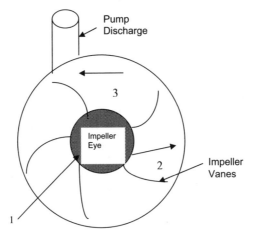

1. Fluid enters the eye of the impeller by pressure flow from the pump suction flange.

2. It is accelerated by centrifugal force to a velocity approaching the tip velocity of the impeller.

3. As the fluid approaches the discharge, the internals of the pump casing allow it to slow down with the inherent increase in pressure.

Figure 7-1   Flow path in centrifugal equipment.

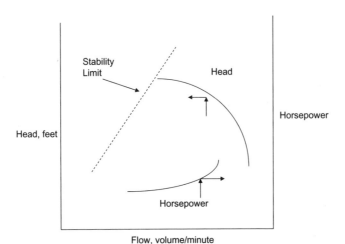

Flow, volume/minute

Figure 7-2   Characteristic centrifugal pump or compressor curve.

occurs, it can damage the equipment. Operation of equipment in parallel can sometimes result in operation within the surging region. This happens because the parallel pieces of equipment have different head curves, or, often, there are slight differences in the mechanical characteristics of "identical" pieces of equipment.

The BEP is the point on the "head curve" at which the hydraulic efficiency is at a maximum. Note that this efficiency does not include motor efficiency. While the head curve is usually developed using water or air, it is valid for any fluid if the correct units are utilized for flow and head. These units are defined later.

As shown in Figure 7-2, horsepower requirements normally peak at the "end of the curve" (maximum flow rate). The driver for the prime mover may or may not be provided with end of the curve protection. If the driver is not designed with end of the curve protection in mind, operating at this point will normally cause the driver to be overloaded. If the driver is a steam turbine, it will slow down, causing the pump to appear to be operating "off the pump curve." If the driver is an electric motor, it will shut down; the electrical load causes the motor protection device to react and shut down the motor before the motor fails due to overload.

7.3 PUMP CALCULATIONS

Before considering exact calculation procedures for pumps, a few definitions are required. These definitions are applicable to positive displacement pumps as well as centrifugal pumps.

- *Cavitation*: A condition that occurs if $NPSH_R > NPSH_A$. If this situation occurs, some of the liquid being pumped will vaporize between the pump suction flange and the pump impeller. This will cause the pump to operate off the head curve, and damage may occur to the impeller.
- *Off the head curve*: This is a condition at which the actual operating point, as defined by the flow rate in gallons/min (gpm), and pressure rise, as described by feet of fluid flowing, is below the pump curve, as defined , for example, in Figure 7-2.
- $NPSH_R$: Net positive head required. This is the head, in feet, required to overcome the pressure loss between the pump suction flange and pump impeller eye. The pump supplier will specify this. A typical $NPSH_R$ versus flow rate curve is shown in Figure 7-3. Note that this curve is usually developed with water, but is valid for any fluid.
- $NPSH_A$: Net positive head available. This is the difference, in feet of head, between the actual pressure at the pump suction flange and the vapor pressure of the liquid being pumped. If the liquid has been stored under a nitrogen, air, or inert gas blanket, some question may arise regarding the actual vapor pressure of the liquid. The most conservative approach is to assume that the vapor pressure is equivalent to the pressure in the storage vessel. If this assumption is used, then the $NPSH_A$ will be the elevation of the drum above the pump impeller less the frictional pressure loss (expressed in feet), assuming there is no change in the suction piping diameter.

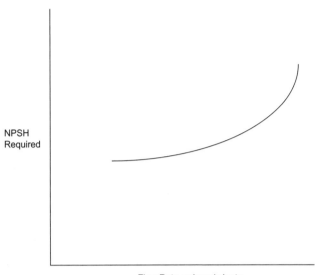

Figure 7-3 Typical NPSH required for centrifugal pump.

The equations below can be used in combination with a pump curve supplied by the manufacturer, similar to Figure 7-2, to determine essential pump operating characteristics. When using the pump curve for a centrifugal pump, the differential pressure across the pump can be calculated as follows:

$$\Delta P = H \times \rho/144 \tag{7-2}$$

where

ΔP = pressure rise across the pump, psi

H = pump head at the given flow rate, ft

ρ = pumped fluid density, lb/ft^3

As indicated earlier, pump curves are usually developed using water as the operative fluid. They are valid for any fluid as long as the density used for calculations is that of the fluid being pumped. Thus in equation (7-2), the pressure developed by a centrifugal pump can be calculated from the head curve supplied by the pump manufacturer, using the head developed at the desired flow rate and the density of the flowing fluid.

The hydraulic horsepower of a pump can be estimated as follows:

$$BHP = H \times F \times 100/(E \times 33,000) \tag{7-3}$$

where

BHP = energy delivered to the fluid, horsepower
F = flow rate through the pump, lb/min
E = hydraulic pump efficiency, %

It should be noted that this calculation only includes the hydraulic efficiency. Other efficiencies (mechanical and/or electrical) must also be considered.

On rare occasions, it will be necessary to determine the temperature change across the pump. For example, pumps with very low flow rate, high recirculation rate from pump discharge to pump suction, and high heads may cause significant heat to be generated and conducted to the fluid recycling to the pump suction line. This higher temperature at the pump suction flange may reduce the NPSH$_A$. The reduction in NPSH$_A$ is due to the increase in vapor pressure as the temperature is increased. The fluid temperature increase across the pump can be estimated as follows:

$$\Delta T = BHP \times (100 - E) \times 2545/(100 \times 60 \times F \times C_P) \qquad (7\text{-}4)$$

where

C_P = fluid specific heat, BTU/lb-°F

This calculation will often indicate that a cooler is required in the fluid recycle line. Even in high head pumps with low flow and no recirculation, heat generation can be a problem. The heat generated by the pump inefficiency will heat up the fluid, which, in turn, will reject some of the heat to the pump case. As the pump case is heated, the NPSH$_A$ will decrease as the vapor pressure at the pump impeller eye increases.

7.4 CENTRIFUGAL COMPRESSOR CALCULATIONS

As indicated earlier, a head curve similar to that shown in Figure 7-2 is valid for centrifugal compressors as well as pumps. The compressor flow is normally expressed as actual cubic feet/minute (ACFM) at the compressor inlet conditions. The ACFM can be calculated by dividing the gas flow rate in lb/min by the gas density in lb/ft^3. Equation (5-3) in Chapter 5 provides the correct approach for estimating the density of a gas. In addition, the calculation of head is much more complicated for a compressor than for a pump. This is because gases are compressible as opposed to noncompressible liquids. The specific volume and density of a liquid, for all practical purposes, can be considered constant. This is not true for gases. As the pressure on a gas is increased, the gas density increases and the specific volume decreases.

While there are multiple references to compression calculations in the literature of this field, two of the references used in the preparation of this book were written by Lyman F. Scheel, *Gas Machinery* and *New Ideas on Centrifugal Compressors, Part 1*. The calculation of head for a compressor depends on the relationship between the pressure and volume of the gas. This relationship is given by the thermodynamic equation:

$$P \times V^k = \text{constant} \qquad (7\text{-}5)$$

where

P = pressure at any point in the compression

V = volume occupied by the gas being compressed at the pressure P

Equation (7-5) determines the exact relationship between pressure and volume. When applying this relationship to compression (centrifugal or reciprocating), the actual value of k depends on which of three thermodynamic assumptions is used. These assumptions are as follows:

1. *Isothermal*: This assumption surmises that the compression occurs at constant temperature. The constant temperature assumption is impossible to achieve in practice.

2. *Adiabatic*: This assumption requires that no heat be added to or removed from the system. This path can be approached in a piston compressor where the discharge and suction compressor valves have exceptionally large areas. The stipulation of exceptionally large area valves is necessary, since the suction and discharge pressures are measured before and after the compressor valves. These compressor valves are flat ribbon valves that are integral to the compressor. However, if the pressures could be measured inside the valves, no stipulation of exceptionally large area suction and discharge valves would be required. In this idealized compression, $k = C_p/C_v$. C_p/C_v is the ratio of the specific heats at constant pressure and constant volume. For an ideal gas, k equals 1.4. If compression is both adiabatic and reversible, it is referred to as isentropic compression. For a compression to be reversible, there must be no friction loss occurring. This type of compression occurs at constant entropy. For this special compression, a Mollier diagram determines the compression path. This description of isentropic compression is included for reference only since, like isothermal compression, it rarely occurs in practice.

3. *Polytropic*: This is an empirical assumption for evaluating the compression path in a compressor. However, it is the normal technique used to evaluate compression head. For polytropic compression, the k value is replaced by n, which is obtained either from the compressor manufacturer or from plant test data. This will be discussed later. The polytropic assumption of compression path is the assumption most frequently used for solving plant operating problems.

The polytropic and adiabatic head can be determined from the same basic equation, shown below:

$$H = 1545 \times T_S \times Z \times (R^\sigma - 1)/M\sigma \qquad (7\text{-}6)$$

where

H = polytropic or adiabatic head, ft

T_S = suction temperature, °R (it should be noted that this suction temperature must be in absolute temperature units of degrees Rankin. This can be determined by adding 460 to the value in degrees Fahrenheit)

Z = average (suction and discharge) compressibility

R = compression ratio (this term is the discharge pressure divided by the suction pressure, in absolute units such as psia)

M = gas molecular weight

σ = polytropic or adiabatic compression exponent

There are three different ways to determine the value of "σ." They are as follows:

1. The compressor manufacturer may supply the value based on test data.
2. The compressor vendor may supply either the polytropic or adiabatic compression efficiency. In this case, the compression exponent (σ) can be calculated as follows:

$$\sigma = (k-1) \times 100/(k \times E) \qquad (7\text{-}7)$$

where

E = either the adiabatic or polytropic compression efficiency, %

k = ratio of specific heats, C_p/C_v. This value is 1.4 for an ideal gas

3. The polytropic compression exponent can be calculated from plant data using the following relationship:

$$T_D = T_S \times R^\sigma \qquad (7\text{-}8)$$

where

T_D = absolute discharge temperature

T_S = absolute suction temperature

It should be recognized that the relationship shown in equation (7-8) gives the compression exponent at any point in time. If the compressor needs mechanical repairs, this value may be higher than predicted by the compressor manufacturer or higher than it was in previously collected plant data. Since the compression exponent calculated from plant data does vary with the mechanical condition of the compressor, this value could be used as a daily monitoring tool for critical compressors. For example, an increase in the compression exponent might indicate that the clearances in a centrifugal compressor had increased to the point at which excessive amounts of gas are recirculating to the compressor suction. An increase in the compression exponent for a reciprocating compressor might indicate that the suction and/or discharge valves should be replaced.

Once the head and flow rate are known, the fluid horsepower requirements for the compressor can be easily calculated, using equation (7-9) below:

$$BHP = F \times H \times 100 / (33,000 \times E) \tag{7-9}$$

where

F = flow rate, lb/min

H = polytropic or adiabatic head, ft

E = either the adiabatic or polytropic compression efficiency, %

A cautionary word is of value at this point. The method used to determine compression efficiency (adiabatic or polytropic) should be consistent with the method used to determine the compression exponent. This will normally be the polytropic efficiency and the polytropic compression exponent. It should also be noted that equation (7-9) includes no mechanical or electrical efficiency that will be associated with belts, gears, or motors.

7.5 DISPLACEMENT SYSTEMS

The term "displacement systems" refers to the prime movers which displace a fixed amount of fluid (liquid or gas), essentially independent of the differential pressure, across the pump or compressor. Typical equipment items that belong within this category are reciprocating pumps/compressors and rotary pumps/compressors. While it might be argued that this type of equipment is no longer used in modern plants, an examination of different processes will show that this type of equipment does indeed have a place in a modern process plant. Examples of the use of this type of equipment are as follows:

- Reciprocating compressors for which the head requirements are high and the volume rate is low to moderate. An example of this is the high-pressure compression step in autoclave and tubular low density polyethylene plants.

- Proportioning pumps with high head requirements or where there is a need to control flow by pump adjustments as opposed to using a flow controller. These might include additive or catalyst pumps.
- Pumps or compressors which must transfer a fixed rate of material regardless of the discharge pressure. This would not be possible with a typical centrifugal system, since the flow rate would decrease as the discharge pressure increased.

There are important concepts which must be understood in order to comprehend this class of prime movers. It should be realized that, with this type of equipment, energy is imparted to the fluid by displacement of a fixed volume of that fluid. The mass flow rate will depend on the fluid suction conditions and physical dimensions of the equipment. That is, larger pump or compressor cylinders will allow displacement of a larger volume of fluid, but the mass of fluid moved also depends on the density at suction conditions.

The calculations described in the previous section are also applicable to this type of system. However, head will not be dependent on flow rate, there will be no head versus capacity curve, and the pump or compressor head will depend on the suction and discharge pressures only. Figure 7-4 shows a typical flow pattern for a reciprocating pump/compressor.

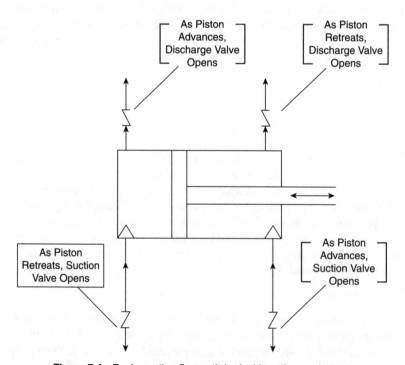

Figure 7-4 Reciprocating flow path in double acting equipment.

Some definitions will aid your ability to evaluate and understand this type of equipment. They are as follows:

- *Volumetric Efficiency*: The actual volume of fluid displaced relative to the volume of the cylinder of a reciprocating pump/compressor or rotating pockets of a rotary pump/compressor. For a liquid, this efficiency approaches 100%. However, for a gas, it is approximately 70%. The differences are due to the compressibility of gases. As the pressure in the cylinder or rotating pocket decreases from discharge pressure to suction pressure, the residual gas in the cylinder/pocket expands to partially fill the cylinder/pocket and reduce the volumetric efficiency of the compressor.
- *Leakage*: This is an additional loss in volumetric efficiency caused by leakage through clearances. An example of this leakage in a reciprocating pump or compressor is the flow of material between the piston and cylinder wall or across the suction and discharge check valves.
- *Horsepower Load Point*: This is a unique feature of a positive displacement compressor. It is the point on a plot of horsepower versus suction pressure at which the required fluid horsepower is at a maximum. On one side of this point, increasing the suction pressure increases the mass flow, which overrides the decrease in compression ratio and causes an increase in the required horsepower. On the other side of this point, the increasing suction pressure results in a decreased compression ratio, which more than compensates for the increase in mass flow rate and results in a decreased horsepower. This concept is discussed in more detail later.
- *Clearance*: This is the part of the cylinder in a reciprocating pump/compressor that is not displaced completely by the piston. Figure 7-5

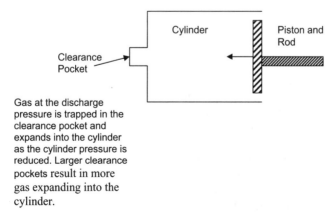

Figure 7-5   Clearance pocket reciprocating compressor.

shows a typical sketch of a "clearance pocket." Because of the compressibility of gases, this becomes more important for gases than for liquids. These are often used to reduce the volumetric efficiency of a compressor and, hence, the required horsepower.

7.6   DISPLACEMENT PUMP CALCULATIONS

Many of the calculations for displacement pumps are identical to those for centrifugal pumps. For example, equations (7-2) and (7-3) can be used to calculate the head/pressure rise relationship and BHP. The main differences in calculations for the two types of pumps are calculation of the fluid flow rate and $NPSH_A$.

For a single-acting reciprocating pump, the following relationship can be used to estimate the volumetric flow rate.

$$V = \Pi \times D^2 \times L_S \times S \times E_V / (4 \times 100) \tag{7-10}$$

where

V = pump capacity, ft^3/min

D = diameter of the pump cylinder, ft

L_S = length of the pump stroke, ft

S = pump speed, RPM

E_V = pump volumetric efficiency, %

There are several points to consider when examining equation (7-10). The pump volumetric capacity is independent of pressure rise and/or pump head, except as the discharge pressure decreases the volumetric efficiency. This is different from the relationship between flow and head for a centrifugal pump.

Equation (7-10) is for a single acting pump, that is, one which displaces the cylinder volume one time for each stroke. The pump that is shown in Figure 7-4 is a double acting pump. It displaces the cylinder two times for each complete stroke. The capacity of a double acting pump will be somewhat less than two times that of a single acting pump of the same dimensions. The actual volume displaced on the return stroke of the piston will be slightly less, due to the volume occupied by the piston and piston rod. Equation (7-10) can be modified for more complicated pumps.

The pump design (reciprocating, plunger, or rotary), the mechanical condition of the check valves, and the internal pump clearances primarily determine the volumetric efficiency. In addition, for high differential pressure pumps, liquid compressibility must also be considered. If the liquid is compressible, the liquid trapped in the clearances will expand when the pressure decreases from discharge pressure to suction pressure. This will cause the volumetric efficiency to decrease. There will also be more leakage across the check valves with high pressure pumps.

The determination of the amount of $NPSH_A$ is more complicated for reciprocating type pumps than it is for centrifugal or rotary pumps. For any type of pump which is pumping a fluid at its boiling point, the $NPSH_A$ is simply the difference in elevation head between the liquid level in the suction drum and the pump suction less the frictional pressure drop. This concept is valid for reciprocating pumps as well, except the frictional pressure drop must be determined at the actual flow rate rather than the average flow rate. The actual flow rate will be greater than the average flow rate due to the cyclic action of the reciprocating pump. In addition, since the fluid in the pump suction line must at some point in the cycle be accelerated from zero velocity to the maximum velocity, there is a pressure loss due to this energy requirement. Fortunately, these two pressure losses occur at different times, so their values are not directly additive. For example, the maximum acceleration head occurs at zero flow rate. Since the flow rate is zero, this is the point of minimum frictional loss. Conversely, the maximum frictional head loss occurs at the maximum rate when there is no acceleration required. For most low-viscosity fluids, the acceleration head dominates. It can be estimated as follows:

1. Calculate the maximum flow rate in the suction pipe.

$$V_P = K \times D^2 \times L_S \times S/(60 \times D_P^2) \qquad (7\text{-}11)$$

where

V_P = peak velocity in suction pipe, ft/sec
D = diameter of the pump cylinder, ft
L_S = length of the pump stroke, ft
S = pump speed, RPM
D_P = diameter of the suction pipe, ft
K = a factor that depends on the pump design

The K value for a double acting pump is approximately $\Pi/2$ or about 1.57. K values for other style of pumps can be obtained from the pump vendors.

2. Calculate the frictional pressure drop, $(h_F)_{MAX}$, in the suction line using conventional techniques, as described in Chapter 5, based on the peak velocity, calculated using equation (7-11).

3. Calculate the acceleration head required to accelerate the liquid in the suction line from zero velocity to peak velocity. This can be determined as follows:

$$(h_A)_{MAX} = 1.35 \times L_P \times S \times V_P / 307 \qquad (7\text{-}12)$$

where

$(h_A)_{MAX}$ = maximum acceleration head loss, ft
L_P = actual suction pipe length, ft
S = pump speed, RPM
V_P = peak velocity in suction pipe from equation (7-11), ft/sec

4. The most conservative approach is to determine the suction piping loss by combining the two types of head loss as follows:

$$(h_T)^2_{MAX} = (h_F)^2_{MAX} + (h_A)^2_{MAX} \qquad (7\text{-}13)$$

where

$(h_T)_{MAX}$ = maximum suction piping head loss to be used to determine the $NPSH_A$

However, as indicated earlier, these two head losses almost always occur at separate times. Thus the use of equation (7-13) may predict overly conservative head losses. The most reasonable estimate of maximum head loss in the suction piping is the larger of the two losses, that is, either the friction head loss or the acceleration head loss.

While equation (7-12) appears to be empirical, it can be easily derived from primary principles of engineering by realizing that the entire mass in the suction line must be accelerated to the maximum suction line velocity as the suction stroke of the pump begins. The value of 1.35 is an experience-based, empirical factor. It is added to allow for the nonsinusoidal motion of the piston and the increased acceleration loss due to elbows and pipe fittings.

Additional information about estimating the $NPSH_A$ for reciprocating pumps can be found in a magazine article entitled *Reciprocating Pumps* by Terry L. Henshaw.

Another aspect that must be considered when dealing with displacement pumps is their "on and off" nature. Rather than the continuous flow that a centrifugal pump provides, the flow will vary from zero to a maximum rate. In some specialized applications, it may be important to assure that there is always some minimal amount of flow from a displacement type of pump. An example of this might be the need to maintain a continuous flow of catalyst to a polymerization reactor. Stopping the flow of catalyst to the reactor, even for a few seconds, might result in plugging of the catalyst injection nozzle. This continuous flow can be obtained in a reciprocating pump system by one of two methods. A more complicated duplex or triplex pump can be utilized. These pumps allow for almost continuous flow of the material being pumped. Another alternative is to install a flow surge bottle in the discharge piping of a simplex reciprocating pump. This vessel, which is equipped with an internal bladder and pressured with nitrogen, can be designed so that the discharge flow to the process approaches an average flow, rather than a peak flow followed by a period of zero flow.

7.7 CALCULATIONS FOR POSITIVE DISPLACEMENT COMPRESSORS

As indicated earlier, the head calculations for both types of compressors are identical. Equations (7-6) through (7-9) can be used to evaluate both head and horsepower requirements for a positive displacement compressor. The actual suction and discharge pressure for a reciprocating compressor may be slightly different than that measured, since the actual pressure is measured in the suction and discharge piping rather than directly at the compressor cylinder during the suction and discharge strokes. In between the compressor cylinder and where the suction and discharge pressures are measured are the ribbon valves. The ribbon valves are pipe-diameter-sized, flat devices which contain multiple strips of metal. These serve as check valves to isolate the compressor cylinder from the suction and discharge pressures at appropriate points in the compressor cycle. For example, when the piston advances, causing a buildup of pressure in the cylinder, the suction ribbon valve closes, keeping gas from flowing back into the suction piping. If the compressor suction and discharge valves are in good mechanical condition, this pressure difference will not be significant. However, if the valves are partially restricted, the pressure at the compressor will be different than that measured. Leakage can also occur in these valves. As indicated earlier, equation (7-8) can be used to monitor the status of these valves for critical compressors.

The flow rate from a positive displacement compressor depends on both the volume of the cylinder or rotating pocket and the volumetric efficiency. As indicated earlier, the volumetric efficiency of a compressor is much lower than that of a pump. This is because gases are more compressible than liquids. The volumetric efficiency of a positive displacement compressor can be estimated as follows:

$$E_V = 100 \times (1 - L - C \times (R^{1/k} - 1)) \qquad (7\text{-}14)$$

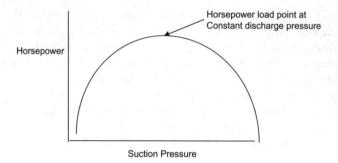

Figure 7-6   Positive displacement compressor load point.

where

E_V = volumetric efficiency, %

C = residual gas remaining in the reciprocating compressor clearance pocket or rotating pocket after discharge, fraction of displacement volume

L = leakage of gas around the piston or rotating element, fraction of displacement volume (roughly 2% of the compression ratio, expressed as a fraction or 0.02)

R = compression ratio

Horsepower load point is a unique feature of a positive displacement compressor. As shown in Figure 7-6, it is the suction pressure at which the BHP reaches a maximum.

To the right of the horsepower load point, the compression ratio and head decrease, but the mass flow increases. To the left of the load point, the compression ratio and head increase while the mass flow decreases. Thus, for a positive displacement compressor, driver overload can occur when the suction pressure is either rising or falling. A single-stage reciprocating compressor, with a constant discharge pressure, constant volumetric efficiency, and an approximate clearance pocket volume of 10%, will reach a maximum horsepower loading at an approximate compression ratio of:

$$R = 1.1 \times (n+1) \qquad (7\text{-}15)$$

where

R = compression ratio

n = constant used to define the polytropic compression exponent $\sigma = (n-1)/n$

Multiple-stage compressors are more complicated, but the concept that decreasing the compression ratio does not always decrease the horsepower requirements is still valid.

7.8   PROBLEM-SOLVING CONSIDERATIONS FOR BOTH SYSTEMS

Compressors

For critical compressors, the suction and discharge temperatures should be used to monitor compressor performance. As indicated previously, equation (7-8) can be used on a daily basis to monitor the performance of a critical centrifugal or positive displacement compressor. If the compression exponent (α) begins to increase, it is an indication that the compressor is becoming less efficient. Another approach that uses the same input is to calculate the "leakage" from the discharge pressure to the suction pressure. An increase in this leakage occurs due to mechanical wear caused by increased clearances in the compressor. This leakage can be calculated as follows:

$$L = [(T_D/(P_D/P_S)^\sigma] - T_S)/(T_D + \Delta T_H - T_S) \qquad (7\text{-}16)$$

where

L = discharge to suction leakage, fraction of volumetric flow

T_D = discharge temperature, °R

T_S = suction temperature, °R

ΔT_H = Joule Thompson cooling effect, °R

P_D = discharge pressure, psia

P_S = suction pressure, psia

σ = compression exponent

The Joule Thompson cooling effect is the amount of temperature change that would be expected as the pressure is lowered from discharge pressure to suction pressure. It is available in various handbooks.

A similar approach can be utilized in monitoring the steam turbines that are often used as drivers on large pumps and compressors. The efficiency of a steam turbine can be determined by comparing the actual change in enthalpy with that predicted assuming an isentropic (constant entropy) expansion. That is:

$$E_T = 100 \times (H_I - H_O)/(H_I - H_E) \qquad (7\text{-}17)$$

where

E_T = turbine efficiency, %

H_I = inlet steam enthalpy, BTU/lb

H_O = outlet steam enthalpy, BTU/lb

H_E = outlet steam enthalpy with isentropic expansion, BTU/lb

The values of enthalpy can be determined from steam charts. These steam charts are available in various publications. While there is a sample problem given later in this chapter, the basic approach for utilization of equation (7-17) is as follows:

1. Knowing the inlet steam temperature and pressure, look up the inlet enthalpy (H_I) and inlet entropy (S_I).
2. Knowing the outlet steam temperature and pressure, look up the outlet enthalpy (H_O).
3. Knowing the inlet entropy and the fact that isentropic expansion (no frictional loss) occurs at constant entropy, look up the constant entropy outlet enthalpy (H_E) based on the outlet steam pressure.

The constant entropy outlet temperature will be different from the actual outlet temperature. It is also possible that this constant entropy expansion may result in an outlet condition that yields a mixture of vapor and liquid. Recognize that this calculated outlet condition is only used so that the turbine efficiency can be calculated. It has no practical use beyond the determination of steam turbine efficiency.

For critical turbines, efficiency can be calculated and monitored on a daily basis. This will allow problem solvers to spot mechanical problems before they become so severe that an immediate repair is required. If problems can be uncovered at an early stage, it is likely that the repair of a steam turbine or compressor can be coordinated with another plant repair downtime.

For centrifugal compressors, the compressor performance curve supplied by the compressor manufacturer can be used to analyze problems. To do this accurately will require careful planning and consideration. The following items should be included in any planned problem-solving activity associated with compressors:

- Make sure that all field instruments have been calibrated before taking any data.
- Determine the kinetic head as described in equation (7-6).
- Calculate the ACFM as accurately as possible. If necessary, adjust the metered flow rate for differences in pressure, temperature, and molecular weight between the meter specification sheet and actual conditions.
- Make sure that the gas composition is known, since it can affect variables such as molecular weight, calculated head, temperature difference between the suction and discharge, and flow rate in lb/hr.

If problem solving involves a plant test on a centrifugal compressor, beware of increasing the speed at constant volumetric flow. Since the surge point normally increases with speed, the compressor could go into surge if the volumetric flow rate is maintained constant when speed is increased.

Similar guidelines are appropriate for positive displacement compressors, except that there will not be a head versus flow curve as there is for a centrifugal compressor. It will be important to have instruments calibrated and to know the gas composition so that the actual capacity can be compared to the predicted capacity. The calculated compression exponent can also be compared to those supplied by the compressor manufacturer.

As indicated earlier, restrictions between the pressure measurement point (discharge or suction) can cause the actual pressures in the compressor cycle to be different from the measured pressures. This could cause the compression exponent calculated from inlet and outlet temperatures to be higher than anticipated. In a reciprocating compressor, the most likely cause of this difference in pressures is ribbon valves that are partially plugged. These valves can also cause a loss of compression capacity if they are partially plugged or if they are leaking. This problem can be monitored and detected using equation (7-8). Chronic valve malfunction is often due to liquid or solid entrainment.

Pumps

One of the most common problems causing poor pump performance is an inadequate available NPSH. While there may be pump manufacturers who claim that their pumps have minimal required NPSH, it should be recognized that all pumps have significant NPSH requirements. The problem solver should look with suspicion on any claims that the NPSH requirements are less than those shown in Table 7-1.

In resolving a centrifugal pump problem, the centrifugal pump performance curve supplied by the pump manufacturer should be used to analyze the problem. If the pump is operating as predicted by the performance curve, then the problem is related to high flow rates or piping limitations. If the pump is not operating on the performance curve, then one of the following five problems may be occurring:

1. The pump clearances may have worn so that a large amount of liquid is recirculating from the discharge to the suction. This will cause the pump to actually be pumping more fluid than the amount shown by flow meters, which will result in a discharge pressure which is lower than anticipated.

Table 7-1 Minimum NPSH requirements

Pump Type	NPSH Required, ft
Centrifugal	6
Centrifugal with booster	1 to 2
Positive displacement	4

2. The $NPSH_A$ is not sufficient. If there is not sufficient NPSH available, some of the liquid will vaporize in the pump inlet, causing the actual amount of liquid being pumped to decrease.
3. A vortex is being formed in the suction side storage tank due to the high velocity in the suction line. The vortex causes vapor to be sucked into the pump suction line. This will cause the pump performance to be below the predicted value, even if the NPSH is sufficient. This problem can be remedied with the installation of a vortex breaker or larger suction line.
4. The pump impeller is the wrong size. Essentially, all centrifugal pumps can be fitted with impellers of different sizes. Often, during routine maintenance, a new impeller of the wrong size is installed in the pump. It should also be noted that most pump curves show multiple impeller sizes. Therefore, before the problem solver can accurately consider a pump problem, he needs to know the size of the impeller that is installed in the pump under consideration.
5. The pump is being driven by a steam turbine and it is not running at the design speed. While this is obviously a steam turbine or steam supply problem, it is often presented to the problem solver as a pump problem.

As indicated for compressors, all instruments should be calibrated and adequately compensated for nondesign conditions before the problem solver begins to collect any data. The problem solver should also be concerned about pumps operating too far from their design point. If a pump is operating at very low flow rates, the resulting instability may cause damage to the pump internals. In addition, high flow can result in excessive $NPSH_R$ or excessive horsepower requirements.

Positive displacement pump problems often are related to either inadequate available NPSH or leaking check valves. The calculation of available NPSH for a reciprocating pump was discussed earlier. Leaking check valves in reciprocating pumps result in a similar loss in capacity as that described for reciprocating compressors.

EXAMPLE PROBLEM 7-1

A centrifugal compressor (C-100L) handling a gas stream that is mostly nitrogen does not seem to have the required process capacity, according to operations personnel. The compressor is driven by a steam turbine. The problem solver associated with this plant was asked to determine the cause of the problem and to recommend actions by which to get the compressor back up to capacity as soon as possible.

Given:

- The compressor curve (Fig. 7-7) is available from the compressor vendor
- Gas molecular weight = 28

Table 7-2   Compressor data

Variable	Run 1	Run 2
Suction pressure, psig	5	5
psia	19.7	19.7
Discharge pressure, psig	28	24
psia	42.7	38.7
Suction temperature, °F	100	100
°R	560	560
Discharge temperature, °F	308	300
°R	768	760
Flow, lb/hr	33,100	41,300

- Gas compressibility $(z) = 1$
- Gas specific heat ratio $(k = C_p/C_v) = 1.4$
- Plant conditions as shown in Table 7-2. All instruments were calibrated before any of the data was taken.

When analyzing the problem, the problem solver used the five-step procedure described earlier. He recognized in the words of the operations personnel that they believed the compressor was not performing as it should, but they had not compared the compressor's actual performance to its predicted performance.

Step 1: Verify that the problem actually occurred.

The problem solver decided that, in order to verify the problem, he had to review the compressor's performance in detail. Therefore, he included the verification step in the problem statement step.

Step 2: Write out an accurate statement of what problem you are trying to solve.

The problem statement that he developed was as follows:

> Plant personnel indicate that the performance of C-100L is not as efficient as they remember it being in the past. The performance of this compressor is causing a decrease in the production limit which does not seem to be easily overcome. Determine if the compressor is performing as predicted by the compressor manufacturer's performance curve. If it is not performing as predicted, determine the cause of the problem and recommend remedial actions.

Note that the above statement emphasizes the step of determining whether a problem exists, compared to the statement provided by operations personnel which emphasizes determining the cause of the problem. Problem-solving

Table 7-3 Compressor calculation steps

1. He calculated α using equation (7-8):

$T_d = T_s \times R^\sigma$ or
$\sigma = (\ln(T_d/T_s)/\ln R)$

	Run 1	Run 2
Suction temperature, °R	560	560
Suction pressure, psia	19.7	19.7
Discharge temperature, °R	768	760
Discharge pressure, psia	42.7	38.7
Compression ratio	2.17	1.96
Compression exponent (σ)	0.408	0.452

2. He calculated the polytropic head using equation (7-6):

$H = 1545 \times T_s \times Z \times (R^\sigma - 1)/M\sigma$

Polytropic head, ft	28110	24400

3. He calculated the ACFM (actual ft^3/min) at C-100L suction:

Gas density, lb/ft^3	0.0919	0.0919
ACFM	6000	7490

4. He looked up the predicted polytropic head from Figure 7-7:

Polytropic head, ft	32000	31500

5. He compared the predicted versus actual polytropic heads, and expressed these values as a percentage:

(Actual − curve) × 100/curve	−12	−23

6. He calculated the polytropic efficiency using equation (7-7) and knowing the compression exponent from step 1 and the specific heat ratio of 1.4:

$\sigma = (k - 1) \times 100/(k \times E)$

$(k - 1)/k$	0.286	0.286
Polytropic efficiency	70	63

7. He calculated fluid horsepower required using equation (7-9):

$BHP = F \times H \times 100/(33000 \times E)$

Horsepower	672	805

steps 3 to 5 cannot be initiated until the alleged poor compressor performance can be verified.

To determine whether C-100L was performing as predicted by the performance curve, the problem solver proceeded and developed Table 7-3.

A reasonable conclusion is that C-100L performs well below the performance curve at the higher flow rate, and performs somewhat below the performance curve at the lower rate. In addition to the poor operation relative to the performance curve, the polytropic efficiency is lower at the higher rates.

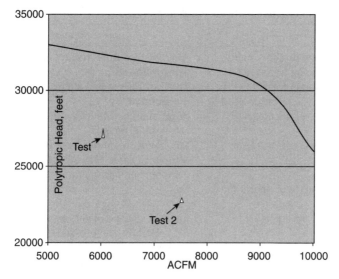

Figure 7-7 Compressor performance curve.

The actual operating points, along with the predicted operating curve, are shown in Figure 7-7. If sufficient data are available, it may be of value to consider when the problem began. The next step is as follows:

Step 3: Develop a theoretically sound working hypothesis that explains as many specifications of the problem as possible.

Referring back to the list of questions given in Chapter 6, the problem solver developed the questions and answers shown below, in Table 7-4.

Note that in formulating answers to the questions, the operator's memory of past performance being better than current performance is not taken as a fact. If the operator's memory was taken as factual, the possibility of design or construction errors could be eliminated. However, memory that is not backed up by hard data is questionable.

Based on the answers to these questions, some possible explanations of the problem were proposed as follows:

1. The suction or discharge piping between the pressure gages and the compressor is too small.
2. There is a restriction in the compressor suction inlet or in the discharge outlet piping.
3. The steam turbine driving the compressor slows as the compressor's horsepower increases.
4. There may be large amounts of compressor internal leakage allowing gases to flow from the discharge to suction side.

Table 7-4 Questions/comments for Problem 7-1

Question	Comment
Are all operating directives and procedures being followed?	All appeared to be correct and are being followed.
Are all instruments correct?	The instruments had allegedly been calibrated.
Are laboratory results correct?	A gas analysis confirmed the molecular weight of the gas.
Were there any errors made in the original design?	The piping might not have been designed for the high rates. However, this would not explain subpar operation at the lower rates.
Were there changes in operating conditions?	No.
Is fluid leakage occurring?	Internal leakage might explain the problem. In addition, the presence of solids might cause a restriction in the suction piping.
Has there been mechanical wear that would explain problem?	Mechanical wear of internals might explain the problem.
Is the reaction rate as anticipated?	Not applicable.
Are there adverse reactions occurring?	Not applicable.
Were there errors made in the construction?	Restrictions in the suction or discharge piping associated with construction might explain the problem.

Hypothesis 3 was eliminated, because the compressor's polytropic efficiency decreases with increasing flow rate. An inadequate steam turbine would not cause the decrease in polytropic efficiency. Hypothesis 4 was treated as a lower priority possibility, even though it would explain a reduced polytropic efficiency. However, it is unlikely that the efficiency would decrease as flow rate increases. The problem solver decided to test both hypotheses 1 and 2 since they are both theoretically sound working hypotheses. Since he was testing both hypotheses, he postponed formalizing the final working hypothesis until he had done additional work.

Step 4: Provide a mechanism to test the hypothesis.

Several methods were developed to test these hypotheses. These tests and the results were as follows:

1. The problem solver calculated the pressure drop in the compressor suction and discharge lines, between the pressure instruments and compressor inlet and outlet flanges. These calculations indicated that the calculated pressure drop was minimal in both lines, thus he decided that hypothesis 1 could be eliminated.

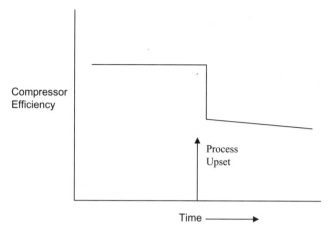

Figure 7-8 Compressor efficiency vs. time.

2. After the pressure drop calculations were made, he arranged for X-rays of the suction and discharge piping to be taken. Based on these X-rays, there appeared to be a significant buildup of solids and/or debris on the bottom of the suction piping before and after the suction pressure instrument.

3. Historical data was reviewed in order to determine the relationship of the compressor's efficiency with time. This relationship is shown in Figure 7-8. It is obvious from this figure that there was a single event in time during which the efficiency dropped dramatically. A review of operating data indicated an upset in the suction knock-out drum that might have caused large amounts of solids to be entrained.

Based on this information, the problem solver developed the following working hypothesis:

The loss of capacity of C-100L appears to be due to the accumulation of solids in the suction piping. This accumulation of solids occurs after the suction pressure gage. The reduced suction pressure is causing the compressor to appear to operate well below the performance curve at high rates, and 12% below the compressor curve at the lower rates. This reduced suction pressure is also causing an apparent loss in compressor efficiency that is more noticeable at high rates. The loss in performance is more noticeable at high rates because the pressure drop caused by the restriction is greater. The presence of solids in the piping may be related to an upset that occurred in the compressor suction drum.

Step 5: Recommend remedial action to eliminate the problem without creating another problem.

Since the X-rays indicated that the likely presence of solids is the most reasonable cause for the loss of compressor capacity, plans were formulated to shut-down the system and clean out the suction piping. While cleaning out the suction piping did eliminate the current compressor limitation, additional problem-solving effort was required. The problem solver next reviewed the upset that appeared to cause solids to be entrained and developed means to prevent this from happening again.

Lessons Learned Problems should always be investigated in as quantitative a fashion as possible. For example, in this problem, the first step that the problem solver took after developing the problem statement was to determine where the compressor was operating relative to an absolute criterion such as the head-flow curve provided by the compressor's manufacturer. It would have been possible to launch into a hypothesis development period before comparing the actual operation to the predicted head curve; however, it would have given equal validity to erroneous hypotheses and valid hypotheses. For example, it was only when the predicted and actual performances were compared at high and low rates that hypotheses 3 and 4 (that there was internal leakage and that the steam turbine was slowing down) could be eliminated.

While hypotheses 1 and 2 were equally valid and could well have been pursued in parallel, it is almost always faster to make calculations to confirm a hypothesis rather than to arrange for elaborate testing. The calculations of pressure drop in the suction piping indicated that the additional cost and time associated with X-raying the suction piping was indeed of value.

This problem also illustrates the flexibility of the five-step approach. The problem solver did not feel that he had sufficient data to propose a hypothesis until he had done some additional work (step 4) including calculations and X-rays. The 5-step approach should not be used to fit all problem-solving activities into a single exact method.

EXAMPLE PROBLEM 7-2

A process that had been idled for 12 months due to a decrease in demand was now required to operate at full rates. A centrifugal pump (P-25) in the process operated as anticipated until the design flow rate was approached. At that point, it no longer operated as predicted by the pump curve. As the unit returned to full capacity, the level in the accumulator drum upstream of the pump had been decreased to provide for more surge capacity at the higher rates. The pump curve is shown in Figure 7-9. The pump was handling a liquid hydrocarbon at its vapor pressure. A schematic flow sheet for the process is shown in Figure 7-10. Operations personnel requested that the problem solver determine why the pump fails to perform at design conditions.

Given:

- The pump curve shown in Figure 7-9
- Plant test data shown in Table 7-5
- Liquid specific gravity = 0.7

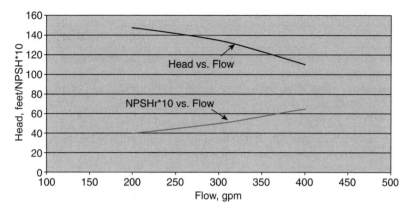

Figure 7-9   Head and NPSHr for Problem 7-2.

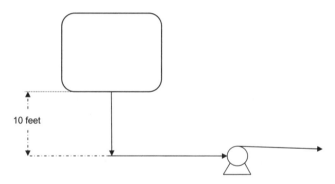

Figure 7-10   Schematic flow for Problem 7-2.

Table 7-5   Plant test data

Test	Run 1	Run 2	Run 3
Flow rate, gpm	200	300	400
Drum pressure, psig	40	40	40
Pump discharge, psig	88	83	65
Drum liquid level, ft	3	3	4

- Hydrocarbon vapor pressure = 40 psig
- Pressure drop in pump suction line = $0.0000528 \times F^{1.8}$, where F = flow rate in gpm

In analyzing the problem, the problem solver used the five-step procedure described earlier. He combined the first two steps as shown below into a problem statement.

Table 7-6   Performance calculations

	Run 1	Run 2	Run 3
1. He calculated the pressure at the pump suction using the relationship given between flow and pressure drop.			
Suction line pressure drop, psi	0.732	1.52	2.55
Elevation head, psi	3.94	3.94	4.25
Suction pressure, psig	43.2	42.4	41.7
2. He calculated the pump differential head using equation (7-2) and knowing the specific gravity relative to water was 0.7 (43.7lb/ft^3).			
Discharge pressure, psig	88	83	65
Differential pressure, psig	44.8	40.6	23.3
Differential head, ft	148	134	77
3. He compared the actual to predicted differential pressure, expressed as a percentage.			
Curve differential head, ft	148	135	110
Error, %	0	0.7	30

Step 1: Verify that the problem actually occurred.

Step 2: Write out an accurate statement of what problem you are trying to solve.

The problem statement that he developed was as follows:

> P-25 has been reported to be operating off the pump curve as the design flow rate is approached. It is likely that this problem was only noticed as the design rates were required after a 12 month period of low rate operation. The level in the accumulator was decreased when the unit was returned to full rates. Since this pump is creating a major production limitation, work to investigate the problem is planned. The first step will be to confirm that P-25 operates on the pump manufacturer's pump curve until the design rate is approached. If the pump does not operate on the curve at all rates, determine the cause of the inadequate performance.

To confirm whether P-25 is operating on the vendor pump curve, the problem solver did the calculations shown in Table 7-6.

The problem solver concluded that P-25 was operating close to the pump curve until the flow rate reached 400 gpm. The next step was:

Step 3: Develop a theoretically sound working hypothesis that explains as many specifications of the problem as possible.

Again, the questions given in Chapter 6 were helpful to the problem solver in formulating possible hypotheses. These questions and the appropriate comments developed by the problem solver for this example problem are shown below, in Table 7-7:

Table 7-7 Questions/comments for Problem 7-2

Question	Comment
Are all operating directives and procedures being followed?	All appeared to be correct and being followed. The operating directive for the level in the drum had recently been reduced.
Are all instruments correct?	The instruments had allegedly been calibrated.
Are laboratory results correct?	Not applicable.
Were there any errors made in the original design?	The piping might not have been designed for the high rates or the increasing $NPSH_R$ as the flow increase might not have been considered.
Were there changes in operating conditions?	No, except for the increased flow and reduced level.
Is fluid leakage occurring?	Internal leakage might explain the problem.
Has there been mechanical wear that would explain problem?	Deterioration of wear rings might explain the problem.
Is the reaction rate as anticipated?	Not applicable.
Are there adverse reactions occurring?	Not applicable.
Were there errors made in the construction?	Not applicable; there had been no recent construction.

Based on this series of questions, two possible hypotheses were developed, as follows:

1. The suction piping is too small or is restricted, causing insufficient NPSH at the pump inlet flange.
2. Deterioration of internal pump wear rings might be causing internal leakage, so that the pump is recirculating large quantities of fluid. This will cause the pump to appear to be operating off the performance curve.

Following the principle of doing calculations prior to making recommendations to do mechanical work, the following calculations shown in Table 7-8 were done.

Since at the highest rate (400 gpm), the $NPSH_R$ (7 ft including the safety factor) is greater than the $NPSH_A$ (5.6 ft) the problem solver developed the following working hypothesis: "The failure of P-25 to operate on the pump curve at high rates is due to the fact that the NPSH is not sufficient. "

Step 4: Provide a mechanism to test the hypothesis.

The problem solver developed two mechanisms to test the hypothesis. The suction line could be replaced, to reduce the pressure drop in the suction line.

Table 7-8 Pump calculations for Problem 7-2

	Run 1	Run 2	Run 3
Vapor pressure of liquid, psig	40	40	40
Suction pressure at pump, psig from Table 7-6	43.2	42.4	41.7
NPSH available, psi	3.2	2.4	1.7
NPSH available, ft	10.5	7.9	5.6
NPSH required, ft			
From curve	4	5	6.5
Safety factor	0.5	0.5	0.5
Total	4.5	5.5	7

A larger suction line that reduced the pressure drop to less than 1 psi would increase the NPSH$_A$ to a value above that required by the pump. Another possible mechanism to test the hypothesis would be to raise the level in the drum. If the level were increased by at least 1.5 ft, the NPSH$_A$ should be adequate.

Step 5: Recommend remedial action to eliminate the problem without creating another problem.

The remedial action depends on which of the hypothesis testing mechanisms is used. If the suction line is replaced and the problem is eliminated, this mechanism becomes the remedial action. If raising the drum level is a successful test, this may or may not be a remedial action. Raising the drum level reduces the amount of surge volume that can be used. A careful study would be required to determine if this action creates other problems.

Lessons Learned   While the available data did not allow the problem solver to eliminate the hypothesis that there was a mechanical problem with the pump that allowed it to recirculate large amounts of fluid, he could use calculations to show that the hypothesis of inadequate available NPSH was a correct hypothesis. Thus if the pump had no mechanical problems at all, it would still not perform as predicted by the pump curve. Therefore, the problem solver elected to develop solutions to the problem of the lack of available NPSH.

Often, in a process unit, operating directives that are changed and seem to work well at reduced rates are not adequate when the rates are increased to design levels. In addition, designs that work at less than design rates may not work at full rates. In the specific example problem, the reduction in the drum level seems to be the cause of the lack of available NPSH. However, additional study would be required before raising the drum level would be considered an acceptable solution.

EXAMPLE PROBLEM 7-3

(This problem is provided to illustrate a calculation technique.) A reciprocating compressor (C-100A) was operating in a mode that resulted in an overload condition on the electric motor. The compressor was equipped with adjustable clearance pockets so that the clearance as a fraction of displacement could be varied. What clearance pocket setting should be used to obtain the maximum flow rate without overloading the electric motor?

Given:

• Clearance pocket setting, % of displacement	4 or 6 or 10
• Suction pressure, psig	15
• Discharge pressure, psig	85
• Compressibility factor (z)	1
• Molecular weight	42
• Polytropic compression exponent	0.32
• Suction temperature, °F	100
• Ratio of specific heats	1.4
• Polytropic efficiency	90
• Mechanical efficiency	95
• Piston displacement, ft³/min	300
• Motor rating, HP	55

The following procedure, shown in Table 7-9, can be used to analyze this problem:

Table 7-9   Calculation procedure

1. Calculate the compression ratio remembering to use psia:

$$R = 99.7/29.7 = 3.36 \tag{7-18}$$

2. Calculate the polytropic head using equation (7-6):

$$H = 1545 \times T_S \times Z \times (R^\sigma - 1)/M\sigma$$
$$H = (1545 \times (100 + 460) \times 1 \times (3.36^{0.32} - 1))/(42 \times 0.32) \tag{7-19}$$
$$= 30500 \text{ ft}$$

3. Calculate the volumetric efficiency for each clearance, using equation (7-14):

$$E_V = 100 \times (1 - L - C \times (R^{1/k} - 1))$$

The leakage factor (L) can be evaluated as 2% of the compression ratio, as follows:

$$L = 2 \times R/100 = 0.07 \tag{7-20}$$

Clearance, %	4	6	10
Clearance factor (C)	0.04	0.06	0.10

(*Continued*)

Table 7-9 (Continued)

Leakage factor (L)	0.07	0.07	0.07
$R^{1/k} - 1$	1.38	1.38	1.38
Volumetric efficiency, %	87	85	79

4. Calculate the mass flow rate and BHP. Use equation (7-9) to calculate the BHP:

$$BHP = F \times H \times 100/(33,000 \times E)$$

Volumetric flow rate, ft³/min	261	255	237
From piston displacement and volumetric efficiency			
Gas density, lb/ft³	0.208	0.208	0.208
Mass flow rate, lb/min	54.3	53	49.3
BHP	58.7	57.3	53.3

Note the BHP includes both mechanical and polytropic efficiency.

Conclusion: The 10% pocket setting must be used to avoid overloading the motor.

EXAMPLE PROBLEM 7-4

The compressor described in Problem 7-3 began losing capacity. The flow meter showed about 35 lb/min. Estimates of motor loading, based on ampere readings, indicated that the compressor required significantly less than 50 BHP. All operating conditions were identical to those specified in Problem 7-3. In addition, the pocket clearance was set at 10%. When the problem solver was asked to consider this problem, he used the five-step approach discussed earlier.

Step 1: Verify that the problem actually occurred.

Verification that the problem was actually occurring was relatively simple, since both the flow meter and the ampere reading indicated that the compressor's output was less than its rated capacity.

Step 2: Write out an accurate statement of what problem you are trying to solve.

The problem solver wrote out the following problem statement:

C-100A seems to be operating well below its design capacity. The flow meter shows 35 lb/hr where the flow should be close to 50 lb/hr. In addition, the horsepower loading is well below the anticipated load of 50 BHP as determined by an ampere reading. The compressor unloading pockets are set so that the clearance is about 10%, which is normal. It is not known when the decrease in capacity actually occurred, but it seems to have been a gradual decrease. Determine what has caused the loss of capacity on C-100A (the reciprocating compressor), why it occurred, and what can be done to eliminate the problem.

Step 3: Develop a theoretically sound working hypothesis that explains as many specifications of the problem as possible.

The questions given in Chapter 6 were used to help formulate possible hypotheses. These questions and the appropriate comments developed by the problem solver for this example problem are shown below, in Table 7-10.

Several possible hypotheses were developed by the problem solver, as follows:

1. The valves in the compressor are bad, causing the loss in capacity.
2. The flow meter is erroneously low.
3. The worn belt drive between the electric motor and compressor is slipping, causing the compressor to operate at a much lower speed than design.
4. The molecular weight of the gas being compressed is different than design.
5. The actual suction pressure is much lower than that shown by the instrumentation.

The problem solver developed a table to attempt to sort out the various hypotheses by comparing the theoretical impact of the hypotheses to the actual observations. This is shown in Table 7-11.

Table 7-10   Questions/comments for Problem 7-4

Question	Comment
Are all operating directives and procedures being followed?	All appeared to be correct and being followed.
Are all instruments correct?	All instruments except the flow meter had been calibrated.
Are laboratory results correct?	Analysis of gas being compressed had not been obtained for several months.
Were there any errors made in the original design?	Not applicable, since the compressor capacity loss was a recent occurrence.
Were there changes in operating conditions?	No.
Is fluid leakage occurring?	Internal leakage might explain problem.
Has there been mechanical wear that would explain problem?	Deterioration of compressor valves might explain the problem. In addition, the belt on the electric motor was beginning to look worn.
Is the reaction rate as anticipated?	Not applicable.
Are there adverse reactions occurring?	Not applicable.
Were there errors made in the construction?	Not applicable.

Table 7-11   Hypothesis comparison

	Would Hypothesis Explain Observations of	
Hypothesis (Number)	Low Ampere Reading	Low Flow Rate
Bad valves (1)	Yes	Yes
Flow meter (2)	No	Yes
Slipping belt (3)	Yes	Yes
Molecular weight (4)	No[1]	?
Low suction pressure (5)	?[2]	Yes

[1]For a reciprocating compressor, the horsepower load is essentially independent of molecular weight. This is true because head is indirectly related to molecular weight (see equation (7-6)), and the mass rate is directly related to the molecular weight. For example, if the molecular weight of a gas being compressed doubles, the polytropic head will be reduced by a factor of 2 and the amount of gas being compressed will increase by a factor of 2. This assumes that the suction and discharge pressures are constant and the only change is in the molecular weight of the gas.

[2]Whether or not a low suction pressure can explain the decrease in horsepower requirements depends upon which side of the load point the compressor is operating at normal conditions. See Figure 7-6 for a typical load versus suction pressure curve. For this specific compressor, it is likely that decreasing the suction pressure will have minimal impact on the horsepower requirements.

Since "slipping belts" (hypothesis 3) will almost always be heard, even above the noise in a compressor house, the problem solver proposed the residual theoretically sound working hypothesis as follows: "The poor performance of C-100A is due to leakage through the suction and/or discharge valves."

Step 4: Provide a mechanism to test the hypothesis.

The actual mechanism for testing the hypothesis involved determination of the "optimum technical depth" required, as described in Chapter 3. Some points that were considered in determining the "optimum technical depth" were as follows:

- How urgent was the problem? Was the loss of capacity causing a loss in plant production, or could gases be diverted to another location?
- Was a spare compressor available? If a spare compressor were available, the required degree of confidence that the solution was correct would be less than it would be if no spare were available.
- Would a compressor shutdown require an entire unit shutdown? If a unit shutdown was required, the required degree of confidence would be much higher.
- Could inlet and outlet temperatures be easily measured, and was historical data available? If these temperatures could be easily obtained and compared to historical data, it would increase the degree of confidence in the solution.

The two possibilities were that the current and historical inlet and outlet temperatures were easily available. In this case, the compression exponent could be calculated using equation (7-8), as shown below:

$$T_D = T_S \times R^\sigma \qquad (7\text{-}8)$$

Then archived data could be used to calculate the historical compression exponent (σ). The current data could be compared to historical data and if a change were obvious, it could be concluded that this was a valid test of the hypothesis. If no historical data is available, equation (7-7) can be used to calculate the theoretical compression exponent from the vendor's performance curve.

$$\sigma = (k-1) \times 100/(k \times E) \qquad (7\text{-}7)$$

This theoretical compression exponent can then be compared to the actual exponent based on plant operating data. If compressor inlet and outlet temperatures are not readily available, it may be possible to use infra-red techniques discussed in Chapter 11 to approximate compressor temperatures.

In this particular problem, a spare compressor was not available. An analysis of the inlet and outlet temperatures indicated that the deterioration of performance had been a gradual decline. This analysis is shown in Table 7-12.

Step 5: Recommend remedial action to eliminate the problem without creating another problem.

Unfortunately, it was necessary to recommend a compressor shutdown to replace the worn valves. Since there was no spare compressor and the com-

Table 7-12   Analysis of compression exponent for Problem 7-4

| Variable | Design | Days After Last Valve Replacement | | | |
		0	30	60	90
Suction					
Temperature, °F	70	70	70	70	70
°R	530	530	530	530	530
Pressure, psig	15	15	15	15	15
psia	29.7	29.7	29.7	29.7	29.7
Discharge					
Temperature, °F	320	320	331	340	360
°R	780	780	791	800	820
Pressure, psig	85	85	85	85	85
psia	99.7	99.7	99.7	99.7	99.7
Compression ratio	3.36	3.36	3.36	3.36	3.36
Compression exponent	0.32	0.32	0.33	0.34	0.36

pressor was an integral part of the unit, it was necessary to shutdown the entire unit. As seems to be the rule, the shutdown occurred at a bad time from a business perspective.

Lessons Learned Replacement of valves on a reciprocating compressor is a "normal maintenance" item. However, since this compressor was in a critical service, the problem solver developed a system to allow inclusion of the compression exponent in a daily monitoring system. His recommendation to do this allowed optimization of the timing for a compressor downtime. In addition, it allowed him to follow the valve performance closely and to potentially correlate valve performance with process upsets or changes that are causing rapid valve wear.

The importance of understanding technology is also illustrated by this problem. The original hypothesis that a change in molecular weight of the gas being compressed was eliminated because a reciprocating compressor's horsepower is essentially independent of molecular weight. In addition, the possibility that a low suction pressure is causing the problem was eliminated since, with this specific reciprocating compressor, suction pressure has minimal affect on horsepower load. These conclusions are not intuitively obvious and can only be developed through an understanding of compression technology.

EXAMPLE PROBLEM 7-5

(This problem is provided to illustrate a calculation technique.) Determine the $NPSH_A$ for a reciprocating pump given the following:

- Pump type   double acting
- Cylinder diameter   0.5 in
- Stroke length   4 in
- Speed   60 RPM
- Suction line diameter   3/4 in
- Suction line length   25 actual ft
- Suction drum pressure   40 psig
- Liquid elevation   10 ft above pump suction
- Fluid vapor pressure   40 psig
- Fluid specific gravity   0.65 relative to water
- Fluid viscosity   0.2 centipoise

As indicated in Section 7.6, there are two kinds of pressure drop in the suction line of a reciprocating pump. There is the typical type of frictional loss, but this occurs at maximum flow rate which is greater than the average flow rate. In addition, the entire volume of fluid in the suction piping must be accelerated from a nonflow condition to the maximum velocity. As indicated in this section, these two frictional losses occur completely out of phase with each other. That is, the frictional loss associated with the maximum flow rate occurs when there is no acceleration. And the acceleration loss from zero velocity to full line velocity occurs when there is no frictional loss. Thus, the pressure loss in the pipe is generally taken as the larger of the two calculated values. The calculations for estimating these pressure losses are based on equations (7-11) and (7-12). The calculation procedure is as follows:

1. Calculate the maximum flow rate in the suction pipe:

$$V_P = K \times D^2 \times L \times S/(60 \times D_P^2) \qquad (7\text{-}11)$$
$$= \pi \times (0.0417^2) \times 0.33 \times 60/(2 \times 60 \times (0.0625^2)) \qquad (7\text{-}21)$$
$$= 0.23 \text{ fps}$$

2. Calculate the frictional pressure drop by conventional means. This calculation is based on equations (5-24) and (5-25) as shown below:

$$\Delta P = 0.323 \times f \times S \times U^2 \times L/d \qquad (5\text{-}24)$$
$$f = -0.001 \times \ln(D \times U \times S/Z) + 0.0086 \qquad (5\text{-}25)$$
$$f = -0.001 \times \ln(0.75 \times 0.23 \times 0.65/0.2) + 0.0086 \qquad (7\text{-}22)$$
$$f = 0.0092$$
$$\Delta P = 0.323 \times 0.0092 \times 0.65 \times (0.23^2) \times 25/0.75 \qquad (7\text{-}23)$$
$$\Delta P = 0.0034 \text{ psi}$$
$$h_F = 0.012 \text{ ft}$$

3. Calculate the head required to accelerate the fluid:

$$h_A = 1.35 \times L_P \times S \times V_P/307 \qquad (7\text{-}12)$$
$$= 1.35 \times 25 \times 60 \times 0.23/307 \qquad (7\text{-}24)$$
$$= 1.5 \text{ ft}$$

4. Select the larger of the two values to determine the actual NPSH for the system.

In this specific problem, the acceleration head will be the critical pressure loss. Actual NPSH at the pump suction will be 10 ft less 1.5 ft, or 8.5 ft. This available NPSH will normally be adequate for most reciprocating pumps.

EXAMPLE PROBLEM 7-6

A centrifugal process gas compressor, C-5A, driven by a steam turbine, no longer had sufficient capacity for the service that it was designed for. The speed of the turbine was controlled by a steam control valve in the incoming steam.

Table 7-13 Design and current operations

Variable	Design	Current Operations
Suction pressure, psig	5	5
Discharge pressure, psig	45	35
Suction temperature, °F	100	100
Discharge temperature, °F	301	264
Flow rate, lb/hr	25000	25000
Gas composition		
Propylene, mol %	95	95
Nitrogen, mol %	5	5
Calculated molecular weight	41.3	41.3
Compressibility	1.0	1.0
Specific heat ratios		
Propylene	1.21	1.21
Nitrogen	1.4	1.4
Mixture	1.2195	1.2195
Turbine speed, RPM		
Maximum	10000	10000
Design	9000	
Actual		8400
Inlet steam conditions		
psig	200	190
Temperature, °F	500	485
Outlet steam conditions		
psig	25	25
Temperature, °F	280	295
Steam flow rate, lb/hr		Out of service
Maximum	16000	
Design	14700	
Steam control valve		
Position, %	80	100

There had allegedly been no changes in operating conditions except for a decrease in steam pressure from 200 psig to 190 psig. This decrease in steam pressure was part of an overall optimization of the plant utility system. In order to increase the steam pressure slightly, operations personnel removed the steam meter orifice that was used to measure the steam flow to the steam turbine. Operations personnel had requested technical help to prove that the steam pressure should be increased back to 200 psig.

The design and current operating conditions are shown in Table 7-13.

The maximum design of the steam turbine included a slight safety factor to allow operations at 10000 RPM with the steam control valve 90% open. As noted in the above table, even with the steam valve 100% open, it is not possible to reach 9000 RPM. The design conditions are those required for the process gas compressor.

The problem solver approached the problem using the 5-step procedure as shown below.

Step 1: Verify that the problem actually occurred.

As the first step in the process of verifying that the problem was real, the problem solver had the appropriate instruments checked. He also used independent sources to confirm the flow and pressure meters. While the compressor output had not been followed on a daily basis, key variables shown in Table 7-13 were available from computer archives. All of these indicated that there was a real problem.

Step 2: Write out an accurate statement of what problem you are trying to solve.

The problem solver wrote out a description of the problem that he was trying to solve as follows:

> Currently, C-5A does not seem to have the desired capacity, even though capacity was adequate in recent history. The steam turbine is operating at a slightly lower speed than that in the design values and the steam control valve is fully open. There has been a reduction in inlet steam pressure from 200 psig to 190 psig. However, it is not clear that this is the cause of the lack of capacity. Determine why C-5A does not have sufficient capacity and provide recommendations for improving the performance of C-5A.

Step 3: Develop a theoretically sound working hypothesis that explains as many specifications of the problem as possible.

The questions given in Chapter 6 were helpful in formulating possible hypotheses. These questions and appropriate comments for this example problem are shown in Table 7-14.

Several hypotheses were proposed as follows:

- When the steam system was optimized, the process designer did not adequately consider the impact of the steam pressure on the steam turbine.
- There is excessive leakage around the wear rings on the process gas compressor. This excessive leakage would cause a compressor efficiency lower than that of the design, as well as a reduced gas rate.
- The steam turbine steam jets may have deteriorated. This would cause the steam turbine to have a lower efficiency and thus extract less horsepower per pound of steam than it was designed to do.

Table 7-14 Questions/comments for Problem 7-6

Question	Comment
Are all operating directives and procedures being followed?	All appeared to be correct and being followed except for the lower steam pressure.
Are all instruments correct?	All instruments except the steam flow meter had been calibrated.
Are laboratory results correct?	Laboratory results indicating that the gas being compressed was 95% propylene and 5% nitrogen were confirmed.
Were there any errors made in the original design?	The original design was okay, but there was a question about whether the change in steam pressure had received adequate consideration.
Were there changes in operating conditions?	Yes. The inlet steam pressure was reduced.
Is fluid leakage occurring?	Internal leakage might explain the problem.
Has there been mechanical wear that would explain problem?	Deterioration of the steam turbine nozzles might explain the problem.
Is the reaction rate as anticipated?	Not applicable.
Are there adverse reactions occurring?	Not applicable.
Were there errors made in the construction?	Not applicable.

Table 7-15 Calculation of polytropic compression exponent

Variable	Design	Operating
Suction pressure, psig	5	5
psia	19.7	19.7
Discharge pressure, psig	45	35
psia	59.7	49.7
Suction temperature, °F	100	100
°R	560	560
Discharge temperature, °F	301	264
°R	761	724
Calculated compression ratio	3.03	2.52
Polytropic compression exponent	0.277	0.278

At this point, the problem solver had developed three theoretically correct working hypotheses, all of which could be evaluated further by additional calculations. Since two of the hypotheses would require compressor shutdowns to inspect and the compressor was not spared, he elected to pursue additional calculations as part of step 4.

Step 4: Provide a mechanism to test the hypothesis.

The problem solver made the following calculations to help determine if one of these proposed hypotheses was both possible and supported by calculations.

1. He determined the polytropic compression exponent using equation (7-8) as shown in Table 7-15:

$$T_D = T_S \times R^\sigma \qquad (7\text{-}8)$$

or

$$\sigma = \ln(T_D/T_S)/\ln R$$

Since the polytropic compression exponent is the same for the design and operating conditions, the problem solver concluded, based on equation (7-7) shown below, that the compressor efficiency was the same as the design efficiency.

$$\sigma = (k-1) \times 100/(k \times E) \qquad (7\text{-}7)$$

2. He calculated the steam turbine efficiency using equation (7-17) and the data in Table 7-13, as shown below:

$$E_T = 100 \times (H_I - H_O)/(H_I - H_E) \qquad (7\text{-}17)$$

The steam turbine efficiency is simply the actual enthalpy change, divided by the enthalpy change at 100% efficiency, multiplied by 100, as shown in equation (7-17) earlier and equation (7-25) below:

$$\begin{aligned} \text{Steam turbine efficiency (design case)} &= 100 \times 91/142 \\ &= 64\% \end{aligned} \qquad (7\text{-}25)$$

It is of value to review how the thermodynamic properties in Table 7-16 were actually developed. The inlet and outlet steam enthalpies were taken from steam enthalpy tables, which are given in multiple publications. These values were used to determine the actual enthalpy change across the turbine. The value shown in the table as "At 100% efficiency" assumes isentropic (constant entropy) steam expansion. The most efficient steam turbine process is one that occurs at constant entropy. The outlet steam conditions are then based on the outlet pressure and maintaining the same entropy as that of the inlet steam conditions. The outlet enthalpy for isentropic expansion is then determined from steam charts or steam tables, as shown in Table 7-17.

The actual steps involved in developing Table 7-17 are as follows:

Table 7-16   Calculation of steam turbine efficiency

Variable	Design	Operating
Inlet steam conditions		
psig	200	190
psia	214.7	204.7
Temperature, °F	500	485
Enthalpy, BTU/lb	1267	1260
Outlet steam conditions		
psig	25	25
psia	39.7	39.7
Temperature, °F	280	308
Enthalpy, BTU/lb	1176	1191
Enthalpy change, BTU/lb		
Actual	91	69
At 100 % efficiency	142	137
Steam turbine efficiency, %	64	50
Calculated steam rate, lb/hr	14700	15800

Table 7-17   Calculation of outlet enthalpy for an isentropic (100% efficiency) expansion

	Design	Operating
Inlet steam conditions		
psig	200	190
psia	214.7	204.7
Temperature, °F	500	485
Enthalpy, BTU/lb	1267	1260
Entropy, BTU/lb-°R	1.615	1.612
Outlet steam conditions		
psig	25	25
psia	39.7	39.7
Entropy, BTU/lb-°R	1.615	1.612
Vapor entropy	1.6763	1.6763
Liquid entropy	0.3919	0.3919
% vapor	95.2	95.0
Vapor enthalpy, BTU/lb	1169.7	1169.7
Liquid enthalpy, BTU/lb	236.03	236.03
Outlet enthalpy, BTU/lb	1124.9	1123.04
Enthalpy change		
Isentropic expansion, BTU/lb	142	137

1. From the inlet steam conditions of temperature and pressure, determine the enthalpy and entropy of the steam.
2. Remembering that the isentropic expansion (100% efficiency) is, by definition, one that occurs when there is no change in entropy, determine the condition of the outlet steam at constant entropy. In this case, the entropy

at 25 psig is between the liquid and vapor entropy. Thus there will be some liquid in the steam leaving the turbine.

3. The actual % vapor can be determined for an isentropic expansion in the design case using the formula below:

$$\% \text{ vapor} = 100 \times (1.615 - 0.3919)/(1.6763 - 0.3919)$$
$$= 95.2\% \tag{7-26}$$

4. Knowing the % vapor, the enthalpy of the outlet steam if the expansion were isentropic can be calculated as shown below:

$$\text{Outlet enthalpy} = 236.03 + 95.2 \times (1169.07 - 236.03)/100$$
$$= 1124.9 \text{ BTU/lb} \tag{7-27}$$

5. The enthalpy change, or work for an isentropic expansion, can then be calculated as follows:

$$\text{Work} = 1267 - 1124.9 = 142.1 = 142 \text{ BTU/lb} \tag{7-28}$$

As shown in Table 7-16, the current efficiency of the steam turbine appeared to be less than design. Thus it seemed unlikely that the current capacity loss was associated with the steam pressure change. Since archived data was available, the problem solver reviewed the historical data to determine when the loss of efficiency occurred and whether it was a onetime loss or a gradual loss. The problem solver used the techniques shown above and developed Figure 7-11. This figure clearly showed that the steam turbine problem was not related to the drop in steam pressure, but was a gradual decay in efficiency that began well after the change in steam pressure occurred.

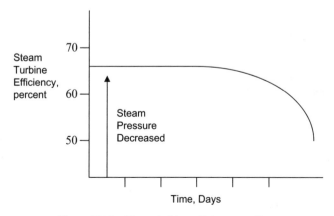

Figure 7-11   Steam turbine efficiency vs. time.

Table 7-18 Compressor horsepower calculations

Variable	Design	Operating
Suction pressure, psig	5	5
psia	19.7	19.7
Discharge pressure, psig	45	35
psia	59.7	49.7
Suction temperature, °F	100	100
°R	560	560
Discharge temperature, °F	301	264
°R	761	724
Flow rate, lb/hr	25000	25000
Polytropic efficiency	65	65
Polytropic compression exponent	0.277	0.278
Calculated polytropic head	27200	22100
Fluid horsepower	530	430

The problem solver had one more question to answer: Would an increase in steam pressure, back to 200 psig, ameliorate the loss of turbine capacity? To answer this question, he calculated the current horsepower load and compared it with the design horsepower load.

Equations 7-6 and 7-9 shown below and the data given in Table 7-13 were used to calculate the polytropic head and fluid horsepower. These equations were then used to develop Table 7-18.

$$H = 1545 \times T_S \times Z \times (R^\sigma - 1)/M\sigma \qquad (7\text{-}6)$$

$$\sigma = (k-1) \times 100/(k \times E) \qquad (7\text{-}7)$$

$$\text{BHP} = F \times H \times 100/(33{,}000 \times E) \qquad (7\text{-}9)$$

He then considered whether the increase in steam pressure from 190 psig to 200 psig would cause the compressor to return to normal. To consider this change in steam pressure, the problem solver first estimated the current steam rate, knowing the delivered fluid horsepower (430 BHP) and the enthalpy change across the turbine. He assumed that steam conditions were returned to the higher pressure, the steam rate increased slightly due to the higher pressure, and that the turbine efficiency remained the same (50%). He then calculated the horsepower that would be delivered to the process, as shown in Table 7-19.

Based on the calculations shown in Table 7-19, the problem solver concluded that raising the steam pressure would not significantly increase the capacity of the process gas compressor. It should be noted that since the turbine efficiency is known, it is not necessary to develop the outlet steam conditions of temperature and enthalpy. However, these conditions could be developed, if desired, using a similar approach as described earlier.

Table 7-19 Calculation of BHP delivered to compressor with increased steam pressure

	Operating Conditions		
Variable	Design	Current	Proposed
Inlet steam conditions			
psig	200	190	200
psia	214.7	204.7	214.7
Temperature. °F	500	485	500
Enthalpy, BTU/lb	1267	1260	1267
Outlet steam conditions			
psig	25	25	25
psia	39.7	39.7	39.7
Temperature, °F	280	308	TBD
Enthalpy, BTU/lb	1176	1191	TBD
Enthalpy change, BTU/lb			
At 100% efficiency	142	137	142
Steam turbine efficiency	64	50	50
Actual	91	69	71
Steam rate	14700	15800	16300
Work from turbine			
MBTU/hr	1.3377	1.09	1.157
Horsepower	525	428	455

He then developed the following hypothesis:

It is believed that the steam turbine steam jets have suffered mechanical damage, which has resulted in the gradual deterioration of their efficiency. This would cause the steam turbine to have a lower efficiency and thus to extract less horsepower per pound of steam than it was designed to do. Increasing the steam pressure from 190 psig to 200 psig would have minimal impact on the turbine.

Step 5: Recommend remedial action to eliminate the problem without creating another problem.

Since all indications were that there was some sort of mechanical damage to the internals of the steam turbine and that the turbine efficiency was continuing to deteriorate, the problem solver had no choice but to recommend that the steam turbine be shut down for repairs. In addition, the time relationship indicated that the repairs should be done as soon as possible. Since issues with safety did not appear to be involved, the actual timing of the shutdown was left to the discretion of the management team.

The management team was pleased with the detailed analysis of the problem because:

1. The initial apparent cause of the problem (the steam pressure reduction) would have been difficult to reverse since it involved several processes in the plant.
2. The relationship of the turbine efficiency with time that was developed was helpful in determining that the steam turbine should be repaired very quickly.

Lessons Learned The value in doing calculations and time related figures is clearly illustrated by this real life problem. Before the problem solver could determine which piece of equipment was the source of the problem, he had to determine the performance of both the compressor and turbine. He determined this by comparing the actual efficiency to the design efficiency. Knowing that the compressor efficiency was essentially the same as the design and that the steam turbine efficiency was less than design allowed him to conclude that the operating problem was related to the steam turbine. In addition, an analysis of the steam turbine supply pressure allowed him to conclude that decreasing the steam pressure was not the cause of the problem and that increasing it back to the original setting would not resolve the performance discrepancy.

If these calculations had not been done, there would be three possible hypotheses that could all have been treated as valid. Without the calculations, the most easily identified change would be the reduction in steam pressure. It would also appear to be the easiest route to improving performance. If the problem solver had not completed the discussed calculations, it is likely that increasing the steam pressure would have been chosen as the route to improve performance. Valuable time that could have been spent in doing a better job in assessing the problem would have been spent raising the steam pressure.

EXAMPLE PROBLEM 7-7

A new process had recently been put into operation. While the startup had gone very well, there was a continuing compressor problem: the main recycle gas compressor would mysteriously shutdown. The 1500-BHP, two-stage recip-rocating compressor was driven by an electric motor. This motor was provided with a shutdown device that was triggered if the horsepower load was 3% greater than the motor rating. Thus the compressor motor would shutdown if the load exceeded 1550 BHP.

The compressor was provided with a "first out" indicator which was used to determine which process variable caused the electric motor to shut down. The first out indicator included process variables such as oil pressure, high discharge temperatures, high suction temperatures, and low suction pressure.

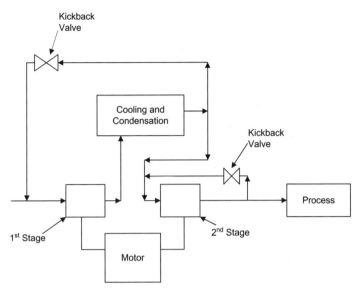

Figure 7-12 Schematic flow for Problem 7-7.

High suction pressure was not included in the first out indicator. However, there was a high suction pressure alarm which would provide a warning prior to the pressure increasing to the point at which the safety valve would release. The indicator panel also had a category called "other reason." Unfortunately, the indicator almost always showed "other reason" for the shutdowns being experienced.

The process gas load was not constant. The gas load varied as the reactor conversion changed; if the reactor conversion decreased, the gas load increased. The process design of the facilities assumed that the conversion would remain constant. Decreases in the conversion were associated with the presence of impurities in the reactor feed or changes in the catalyst flow rate to the reactor.

The compressor was provided with computer-controlled unloaders and recycle capabilities to maintain a constant suction pressure on both stages. If the gas load was such that both the recycle valves were closed and the unloaders were in a position to compress the maximum amount of gas, the suction pressure would increase until the gas density increased to the point that the compressor had adequate capacity. A schematic sketch of the facilities is shown in Figure 7-12. The key variables for the compressor are given in Table 7-19.

Operations personnel have requested help from the technical organization to determine what was wrong with the first out indicator. The problem solver approached the problem using the 5-step procedure, as shown below.

Table 7-20   Compressor operating data for Problem 7-7

First-stage suction pressure, psig	3
Second-stage suction pressure, psig	60
Interstage pressure drop, psi	2
Discharge pressure, psig	275
Compressibility factor (z)	1
Molecular weight	42
First-stage suction temperature, °F	100
First-stage discharge temperature, °F	283
Second-stage suction temperature, °F	110
Second-stage discharge temperature, °F	280
Interstage condensation, lb/hr	0
Ratio of specific heats	1.21
Polytropic efficiency	90
Mechanical efficiency	95
Piston displacement, ft³/min	5000
Motor rating, HP	1550 (includes a 3% overload factor)

Step 1: Verify that the problem actually occurred.

There was no doubt that the compressor was shutting down. However, it wasn't obvious that there was anything wrong with the "first out" indicator. There were times when it showed indications besides "other reason." In addition, simulated signals were used when the system was out of service to confirm that all other indicators worked. The problem solver expanded the original problem scope and stated it as shown in step 2.

Step 2: Write out an accurate statement of what problem you are trying to solve.

The statement written by the problem solver was as follows:

> C-122 has been shutting down for no apparent reason. The "first out" indicator shows the cause is "other reason." At about the same time as the mysterious shutdowns, the suction pressure tends to increase. However, it has been impossible to determine whether this causes the shutdowns or is the result of the shutdowns. Investigate the shutdowns of C-122, the recycle gas compressor, to determine what is causing the unknown shutdowns. There may be a problem with the first out indicator or there may be an unknown reason for the compressor shutdowns. When the cause is determined, provide recommendations to eliminate the problem.

Step 3: Develop a theoretically sound working hypothesis that explains as many specifications of the problem as possible.

The questions given in Chapter 5 were used to help formulate possible hypotheses. These questions and appropriate comments for this example problem are shown below, in Table 7-21:

Table 7-21   Questions/comments for Problem 7-7

Question	Comment
Are all operating directives and procedures being followed?	All appeared to be correct and being followed, though there were times when the low-stage suction pressure was not well controlled.
Are all instruments correct?	All pressure instruments had been calibrated.
Are laboratory results correct?	Not applicable.
Were there any errors made in design?	Since this was a new process, this was a consideration. However, the major concerns were the first out indicator and the capability of the compressor to handle swings in recycle gas rates with subsequent changes in suction pressure.
Were there changes in operating conditions?	No, except for the suction pressure swings.
Is fluid leakage occurring?	Internal leakage might explain the failure of the compressor to handle the gas flow under some conditions.
Has there been mechanical wear That would explain problem?	Leaking compressor valves might explain the problem.
Is the reaction rate as anticipated?	Normally yes. However, decreases in reaction rate cause increased recycle gas rates.
Are there adverse reactions occurring?	Not applicable.
Were there errors made in the construction?	Construction errors in the first out system might explain the problem.

Based on the above table, the problem solver formulated the following hypotheses.

1. There is a compressor mechanical problem, such as bad valves. This problem is not apparent at low flow rates, but becomes apparent at high rates. This mechanical problem causes the compressor to have insufficient capacity at high gas rates. As the suction pressure increases, the compressor shuts down for some unknown reason.
2. The compressor is shutting down, for some unknown reason, when the suction pressure rises slightly due to normal process variability.
3. There is an intermittent instrumentation failure with the first out system. This failure is causing some problem, such as low oil pressure, to shutdown the compressor but not show up on the display panel.

Note that there is similarity between the first and second hypothesis. The first hypothesis implies that there is a mechanical condition that can be repaired

Table 7-22   Calculations of compressor efficiency

Variable	Design	Current
Low-stage suction pressure, psig	3	3
psia	17.7	17.7
Low-stage discharge pressure, psig	62	62
psia	76.7	76.7
High-stage suction pressure, psig	60	60
psia	74.7	74.7
High-stage discharge pressure, psig	275	275
psia	289.7	289.7
Low-stage suction temperature, °F	100	100
°R	560	560
Low-stage discharge temperature, °F	278	283
°R	738	743
High-stage suction temperature, °F	110	110
°R	570	570
High-stage discharge temperature, °F	276	280
°R	736	740
Compression exponent (from temperatures)		0.193
Polytropic efficiency, %	92	90
Polytropic head, ft	67200	67400
Mass flow rate, lb/min	620	620
Total horsepower, BHP	1440	1480

and the problem will be eliminated. The second hypothesis indicates that the mysterious shutdowns are due to process variability and that there is nothing mechanically wrong with the compressor.

When faced with the need to decide which of these hypotheses to pursue, the problem solver recognized that it might be difficult to trace an intermittent instrument failure associated with the first out system. On the other hand, it would be possible to quickly do some calculations to confirm whether or not either hypotheses 1 or 2 were theoretically correct. He used the following relationships and the basic data shown in Table 7-20 to develop Table 7-22.

$$T_D = T_S \times R^\sigma \qquad (7\text{-}8)$$

The problem solver used equation (7-8) with the suction and discharge temperatures in absolute temperature units and calculated a compression exponent of 0.193. He then used the compression exponent (σ), the specific heat ratio, and equation (7-7), shown below, to calculate the polytropic efficiency. The ratio of specific heats is given in Table 7-20.

$$\sigma = (k-1) \times 100/(k \times E) \qquad (7\text{-}7)$$

Table 7-23   Calculated horsepower load versus suction pressure

Variable	Case 1	Case 2	Case 3
LS suction pressure, psig	3	4	5
psia	17.7	18.7	19.7
LS discharge pressure, psig	62	62	62
psia	76.7	76.7	76.7
HS suction pressure, psig	60	60	60
psia	74.7	74.7	74.7
HS discharge pressure, psig	275	275	275
psia	289.7	289.7	289.7
LS suction temperature, °F	100	100	100
°R	560	560	560
LS discharge temperature, °F	283	275	267
°R	743	735	727
HS suction temperature, °F	110	110	110
°R	570	570	570
HS discharge temperature, °F	280	280	280
°R	740	740	740
Compression exponent	0.193	0.193	0.193
Polytropic efficiency	90	90	90
Polytropic head	67400	65900	64500
Mass flow rate, lb/min	620	655	690
Total horsepower	1480	1523	1575

As shown in Table 7-22, the polytropic efficiency is slightly lower than the design efficiency (90 vs. 92). This slightly low efficiency is likely within the accuracy of the calculations. The mass flow and calculated total horsepower are also shown. The problem solver concluded that the hypothesis that the compressor required mechanical repairs was not a valid hypothesis. He then considered the question would process upsets such as a sudden increase in the recycle gas rate be sufficient to overload the compressor. He developed a spreadsheet to allow calculation of the horsepower load at various low-stage suction pressures. The results of these calculations are shown in Table 7-23.

The problem solver believed that his calculations clearly indicated that the reason for the mysterious compressor shutdowns was associated with a slight increase in process gas rates that caused the compressor suction to increase to the point at which the compressor motor was overloaded. This was in spite of the intuitive thought process that implied that the horsepower load should decrease as the compression ratio decreases. The 1500 BHP motor with a 3% overload factor would likely shutdown if the compressor suction pressure exceeded 4.5 psig. He expressed his hypothesis as follows:

It is believed that the mysterious recycle compressor shutdowns are caused by spikes in the recycle gas rate that cause the motor to overload as the suction pressure is increased.

While it might seem that this hypothesis could be confirmed by a simple analysis of the low-stage suction pressure and compressor amperes versus time, the exact point of shutdown was not obvious. As the compressor suction pressure rose and the compressor shutdown, the pressure continued to rise rapidly. It was impossible to determine whether the shutdown occurred when the pressure reached 4, 4.5, 5, or 6 psig.

Step 4: Provide a mechanism to test the hypothesis.

The problem solver considered that the best option to test his hypothesis was to run a test at reduced production rates. A test was run at a low enough production rate that the recycle gas never got high enough to cause the suction pressure to go above 4 psig. During this test, there were no mysterious compressor shutdowns.

Step 5: Recommend remedial action to eliminate the problem without creating another problem.

While this test was successful at preventing the mysterious compressor shutdowns, it was obviously not a permanent solution. Experience with the process over the early startup period indicated that the recycle gas rate was likely to increase to 15% above the steady state design rate during decreases in reactor conversion. The problem solver looked at an increase of 15% in gas rate and concluded that the suction pressure would increase from 3 psig to 5.7 psig if this occurred. At 5.7 psig, the calculated horsepower load would be 1610 BHP. This would be well above the motor acceptable load with its 3% overload factor. He discussed the situation with the motor manufacturer and found out that operating the motor continuously at 10% overload would be expected to shorten the life of the motor by a slight amount. Since the 10% overload would only occur during times when the reactor conversion dropped, a larger overload switch was installed.

Lessons Learned   While this problem solution may seem obvious, it should be recognized that the initial assessment was that there was a problem with the first out instrumentation system. If this idea had been followed through with no consideration of other possible hypotheses, there would have been a significant delay in solving the problem since the problem was thought to be an intermittent failure. In addition, while the 3% overload rating may seem too conservative, without the calculations done by the problem solver it would not be known whether the 10% overload rating would cover the range of loads to be expected. The calculations also helped to steer the problem solver away from the conclusion that there was a mechanical problem with the compressor.

This problem also illustrated that what might seem to be intuitively correct (lowering the compression ratio by increasing the suction pressure will decrease the horsepower load) is not always true. While it is true that lowering the compression ratio by increasing the suction pressure lowers the polytropic head, it also increases the mass flow rate. Both polytropic head and mass flow rate are important in determining the horsepower load. The exact relationship between suction pressure and horsepower load can only be determined by calculations.

NOMENCLATURE

BHP	Energy delivered to the fluid, horsepower
C	Residual gas remaining in the positive displacement compressor clearance pocket or rotating pocket after discharge, fraction of displacement volume
C_P	Fluid specific heat, BTU/lb-F°
D	Diameter of the pump cylinder, ft
D_P	Diameter of the suction pipe, ft
E	Hydraulic pump efficiency or adiabatic/polytropic compression efficiency, %
E_T	Turbine efficiency, %
E_V	Pump or compressor volumetric efficiency, %
F	Flow rate, lb/min
g_c	Gravitational constant
H	Pump head or adiabatic/polytropic head, ft
H_I	Inlet steam enthalpy, BTU/lb
H_O	Outlet steam enthalpy, BTU/lb
H_E	Outlet steam enthalpy with isentropic expansion, BTU/lb
$(h_A)_{MAX}$	Maximum acceleration head loss for a reciprocating pump, ft
K	A factor for reciprocating pumps that depends on the pump design
k	Ratio of specific heats, C_p/C_v
L	Leakage of gas around the piston or rotating element, fraction of displacement volume. It is roughly 2% of the compression ratio expressed as a fraction
L_S	Length of the pump stroke, ft
L_P	Actual suction pipe length, ft
lw	Frictional loss in the piping system
M	Gas molecular weight
n	Constant used to define the polytropic compression exponent σ= $(n-1)/n$

P_D	Discharge pressure, psia
P_S	Suction pressure, psia
R	Compression ratio
S	Pump speed, RPM
T_D	Absolute discharge temperature, °R
T_S	Absolute suction temperature, °R
V	Reciprocating pump capacity, ft³/min
V_P	Peak velocity in suction pipe of a reciprocating pump, ft/sec
w	Amount of work added by the prime mover
Z	Average (suction and discharge) compressibility
ΔP	Pressure difference between two points or pressure rise across a pump
ΔT_H	Joule Thompson cooling effect, °R
$\Delta(v^2)$	The difference in velocity squared between two points
Δz	Difference in elevation between two points
ρ	Pumped fluid density, lb/ft³
σ	Polytropic or adiabatic compression exponent

8

APPLICATION TO
PLATE PROCESSES

8.1 INTRODUCTION

Examples of plate processes are fractionation towers, extraction towers, absorption towers, or any process that depends on mechanical design to provide intimate contact between two phases followed by a zone where phase separation is achieved. The intimate contacting followed by phase separation allows equilibrium between the two phases to be achieved or approached.

The information provided in this chapter is intended to provide a basis for the problem solver to successfully complete step 3 of the problem-solving discipline: "Develop a theoretically sound working hypothesis that explains as many specifications of the problem as possible."

Since much of what occurs in a plate process is not visible to the naked eye and may not be easily understood by X-rays, it is important to be able to correctly imagine what is occurring inside the equipment.

Fractionation using sieve trays is discussed in this chapter since it is the most common application. However, the principles are applicable to all fractionation tray designs and any other plate processes.

8.2 FRACTIONATION WITH SIEVE TRAYS

The purpose of a sieve tray is to provide as close an approach to equilibrium between the liquid and vapor phase as is reasonably possible. The concept of vapor-liquid equilibrium was discussed in Chapter 5. To obtain or approach equilibrium, the following three zones are required:

Problem Solving for Process Operators and Specialists, First Edition. Joseph M. Bonem.
© 2011 John Wiley & Sons, Inc. Published 2011 by John Wiley & Sons Inc.

1. *A high intensity vapor-liquid contact zone.* In this zone, liquid must be the continuous phase. That is, vapor bubbles must exist as discrete entities surrounded by a liquid phase regime. Vapor will be bubbling up through the liquid. Due to the energy being expended, the liquid will appear as a frothy mixture of liquid and vapor.

2. *An entrained liquid separation zone.* This vapor-continuous zone is immediately above the continuous liquid phase. In this zone, the liquid droplets entrained with the rising vapor disengage and return to the liquid phase under the influence of gravity.

3. *A vapor-froth separation zone.* The liquid leaving each tray contains vapor that has not yet disengaged from the liquid. The downcomer provides time for disengagement to occur. This will result in vapor-free liquid exiting the downcomer on the tray below.

Figure 8-1 shows a typical sieve tray illustrating these three zones. In addition, it indicates two other important parameters for a fractionating tray design. These are as follows:

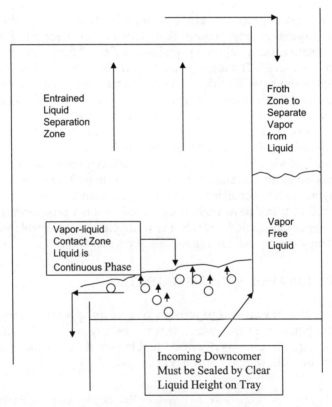

Figure 8-1 Typical fractionation tray.

1. The inlet downcomer must be sealed to prevent vapor from going up the downcomer. The sealing is accomplished by insuring that the pressure head of the clear liquid on the tray is greater than the clearance between the downcomer and the tray deck. The clear liquid head is the height the liquid would attain if there were no vapor present bubbling through the liquid. This sealing criterion will force liquid to build up in the downcomer so that the downcomer is sealed. If this criterion is not met, the level in the downcomer will be minimal. This will allow vapor to flow up the downcomer rather than through the holes on the tray above.

2. The hydraulic gradient (the difference in clear liquid height between the inlet downcomer and outlet downcomer) must be minimized. If the hydraulic gradient is too great, the trays will have a tendency to weep liquid on the inlet side of the tray. Weeping is a tendency of liquid to flow down through the holes on the tray rather than flowing through the downcomer.

Tray stability diagrams provide an analytical means to help visualize what is occurring on a fractionating tray. Figure 8-2 provides an example of a tray stability diagram. The four areas of unacceptable operation are:

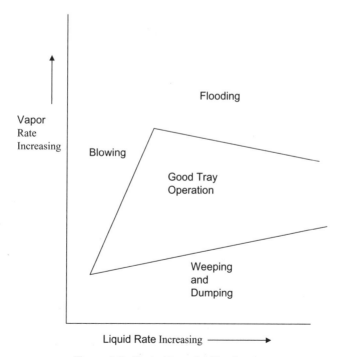

Figure 8-2 Typical tray stability diagram.

1. *Flooding*: This condition is marked by vapor velocities based on the tower cross-sectional area being so high that large amounts of liquid are carried up into the continuous vapor phase regime. This liquid does not adequately disengage from the vapor and is carried up into the tray above. This condition leads to excessive loading in the tray outlet downcomer as the entrained liquid returns to the tray below via the downcomer. This excessive loading causes the liquid level in the downcomer to build up to an unacceptable level. Flooding tends to start at one tray and propagate upward through a section of the tower. It may often be detected by a measurement of differential pressure across a section of the tower.

2. *Downcomer Filling*: This condition is marked by the downcomer either completely filling or filling to the point that adequate vapor disengagement from the liquid cannot occur. It can be caused by flooding, as indicated earlier, or by excessive liquid rates. It can also be caused by tray or downcomer restrictions. In a similar fashion to flooding, it can propagate upward through each tray in a fractionating tower.

3. *Blowing*: In this condition, vapor velocity through the holes in a fractionating tray is so high that the vapor phase becomes the continuous phase in the high intensity vapor-liquid contact zone. If the fractionating column were made of clear material, this condition would be observed as one in which the liquid is blown off the trays. It is usually caused by a combination of low liquid rates and high vapor rates.

4. *Weeping/Dumping*: In this condition, the vapor rates are so low that liquid pours down the holes in the tray. This results in a very low liquid level on the trays. This usually results in the loss of the downcomer seal, causing vapor to flow up the downcomer. This single unsealed downcomer will likely cause a high degree of frothing, which will lead to liquid holdup in the downcomer. This may result in downcomer flooding in the trays above the tray that is weeping and/or dumping. It should be noted that tray weeping/dumping could also be caused by mechanical damage such as a tray segment that has come loose and is hanging down from its support.

As a general rule, the symptoms of "flooding" and "downcomer filling" are a higher-than-normal pressure drop across the tower and loss of fractionation. The "blowing" and "weeping/dumping" symptom is loss of fractionation in the tower. Pressure drop for these conditions may be normal or slightly lower than normal.

The tray stability diagram can be developed by the following procedure.

1. Assume a liquid rate.
2. Vary the assumed vapor rate at this assumed liquid rate until the weeping, blowing, flooding, or downcomer filling limits are encountered.
3. Repeat this calculation for several different liquid rates until the tray stability diagram can be completed.

The calculation techniques for determining the various tray limitations are beyond the scope of this book. Adequate references are available in the open literature within this field or in individual company literature. The key point is that a tray stability diagram provides a means to visualize what is happening in the tower and thus will be helpful in steps 3–5 of the disciplined problem-solving approach.

8.3 PROBLEM-SOLVING CONSIDERATIONS FOR FRACTIONATING TOWERS

Assuming that the fractionating tower was designed correctly, problems will almost always be associated with process changes or mechanical damage. Confirmation that the fractionating tower was designed correctly can be attained through development of a tray stability diagram. This tray stability diagram will allow operators to plot operating conditions on the diagram to confirm that they are within the specifications of "good operations." Developing this tray stability diagram is also consistent with the premise of this book: Calculations should be made prior to developing hypotheses that might explain the problems being encountered.

Mechanical damage to trays can create tray performance problems. Some of these possible areas of tray damage or improper tray installations are described in the next few paragraphs.

A tray segment can fail. Most sieve trays are designed to be installed and/ or removed through manways that have a diameter of 18 to 36 in. This approach also facilitates tray inspection and replacement. Thus a single tray will consist of several segments connected together. If one of these connections should fail, a condition similar to weeping/dumping will occur. As indicated earlier, this can propagate upward, causing a potential loss of several trays.

The trays may not be level. While this is almost always an installation problem, it might also occur after an extended period of operation. This delayed manifestation could be due to extreme foundation settling. In addition, problems associated with uneven trays might not show up under all operating conditions. A tray slope of less than 0.2% is generally acceptable and slopes of up to 0.7% have provided good operations. Slope is the amount of elevation change per inch of tray diameter expressed as a percentage. Thus an elevation change of 0.5 in in a 6-ft diameter tower is equivalent to a slope of 0.7%. Highly critical towers (e.g., vacuum towers) are special cases.

There may be restrictions in the tray. The presence of solids in a fractioning tower will often lead to plugging of a downcomer or plugging of holes in sieve trays. Solids can be present in the tower due to various causes, such as:

- *Construction debris*: It is not unheard of to discover items such as rags, tools, or even safety equipment left behind in a tower after construction or repairs. Unfortunately, there have been occasions on which an inspection of the tower prior to startup did not find the debris which blocked a downcomer. Following the startup of the tower, problems were experienced and a subsequent shutdown and inspection revealed the debris in the downcomer.

- *Entrainment from a drum containing solids*: Very fine particles can often be entrained from a vapor-liquid disengaging drum and carried into a tower. If these particles are not soluble in the liquid on the trays in the tower, they can accumulate in either the downcomer or sieve tray holes.

- *Corrosion of the tower internals*: If the liquid on the trays contains a corrosive compound, the trays may begin to corrode. This could cause the holes in the sieve trays to enlarge. In addition, it is possible that some of the corrosion products could be deposited lower in the tower as the composition of the liquid changes. Because of this possibility, sieve trays are often fabricated from a corrosion-resistant material, such as stainless steel.

- *Reaction of trace components inside the tower*: While this would seem like an unlikely event, there have been instances during which small quantities of water reacted with a soluble material in the liquid phase on the tower trays and formed an insoluble material. This material would accumulate in the sieve tray holes and the downcomers and create plugging problems which manifest similarly to the item discussed earlier.

The vapor inlet flow pattern may change. For example, the holes in a sieve tray may enlarge or become irregular due to corrosion, or the valves in a valve tray may be loosen. Either of these occurrences may cause significant disturbances to the vapor inlet flow pattern and result in poor tray performance.

Process changes causing poor tray performance may be due to known changes, such as an increase in rates, or very subtle changes. For example, there may be instrumentation errors that cause excessive vapor or liquid rates. Even at constant liquid and vapor rates, foaming caused by trace quantities of a surface active ingredient may occur. An even more subtle change might be the tower that appears to be operating perfectly normally until an event such as that described earlier. The presence of water from an exchanger leaks reacts with a soluble material to form solids. These solids lead to plugging of the holes in the sieve tray. The first indication of such a problem might be the increased differential pressure across the tower.

8.4 DEVELOPMENT OF THEORETICALLY SOUND WORKING HYPOTHESES

Once step 1 (verify that the problem actually occurred) and step 2 (write out an accurate specification of the problem) have been completed, the following guidelines can be used to develop a working hypothesis.

Calculations can be performed to determine the following:

- *Tray stability diagram*: This will highlight the areas of unacceptable tray operation as well as indicate the current point of operation. If the current operations are in an unacceptable area, this must be corrected before any other calculations are considered.
- *Number of theoretical stages required for the degree of separation being encountered*: Knowing the concentrations at various points in the tower, the number of theoretical stages required to make this separation can be estimated by computer or manual techniques. These manual techniques and their place in the modern world are discussed in Chapter 13.
- *Estimated tray efficiency*: The tray efficiency is the theoretical stages divided by the actual stages and expressed as a percentage. It is a technique to allow one to estimate the number of actual stages required if the number of theoretical stages is known. While this book does not cover these techniques, there are multiple methods available to estimate the tray efficiency.

While these calculations are beyond the scope of this book and would normally be performed by a graduate engineer, the process operator serving as a problem solver needs to be aware of them. These calculations are important in assessing and answering the question of "Was the tower designed correctly and is it operating correctly?" If the process operator is aware of what kind of calculations can be performed, he will be able to develop a better plan of action.

It should be noted that these calculations will require accurate plant data and tower/tray design information. Instrumentation should be calibrated before using plant data to perform these calculations, to avoid having to redo the calculations.

Assuming that the calculations described above indicate that the tower should be performing better than what the actual data indicates, the list of questions given in Chapter 6 can be used to help formulate a working hypothesis. Once a hypothesis is formulated, it may be of value to develop a plant test to confirm the hypothesis. Measurements of tray temperatures, pressures, and compositions will be helpful in determining what areas of the tower are worthy of future analysis. Several meaningful approaches are shown in Table 8-1. In addition, x-rays can be utilized to examine the suspect area of the tower either to help formulate a hypothesis or to provide a mechanism to test the hypothesis (step 4: provide a mechanism to test the hypothesis). Some examples of the use of x-rays are shown in Figure 8-3.

8.5 PROBLEM SOLVING AND REBOILER CIRCUITS

Reboilers, which are often straightforward in the conceptual stage, are often a significant part of a plant fractionation problem. The most frequent culprit

Table 8-1 Evaluation of trays by sampling/data analysis

Measurement	Typical Values	Meaning of Atypical Values
Pressure drop across trays	Should be measured across as few trays as possible Check measurements with calculations Normal values 0.05–0.20 psi/tray. Lower for vacuum towers	Trays plugged, damaged or flooded
Temperature change across trays	Should be equivalent to that estimated from fractionation calc. or equivalent to historical values.	Tray damage or process changes
Composition of liquid on trays	Should be equivalent to that estimated from fractionation calc.	Numerous
Venting sample bomb of tray liquid	For pressure towers, vaporizing liquid should cool off bomb	Liquid phase is not present on tray
Composition of vapor on trays	Should be equivalent to that estimated from fractionation calc.	Numerous
Venting sample bomb of tray vapor	No temperature change	Tray flooding

is the vertical thermosiphon reboiler. A typical flow sheet is shown in Figure 8-4.

Thermosiphon reboiler problems are often due to hydraulics. Since these reboilers do not have a pump associated with them, circulation depends on the hydraulic balance around the tower bottoms and reboilers. The operation of this class of reboilers depends on a delicate balance of elevation, fluid densities, and pressure drop. The density of the return line to the tower is a function of the percentage vaporization. These thermosiphon reboilers almost never operate with pure vapor in the return line to the tower. Thus the amount of vaporization (percent vaporization) in the reboiler will determine the density in this tower return line. There must be sufficient head to cause the system to circulate with no application of external work. Some of the possible causes of hydraulic problems are described in the following paragraphs.

There may be inadequate elevation head of liquid in the tower bottom. If the liquid level in the bottom of the tower is not high enough, the driving force to cause the process liquid to flow will be inadequate. This may result in a lower-than-design liquid level in the reboiler and/or a lower-than-design circulation rate. The low reboiler level will cause the surface area available for vaporization to decrease with a likely loss of heat input to the tower. In addition, if the process liquid flow decreases and the heat input remains the same, the percentage vaporization will increase. In some applications, this may create reboiler fouling.

1. Dropped Tray Segment

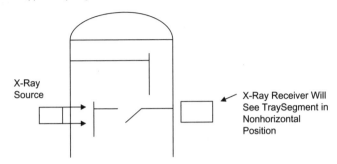

2. Plugged Downcomer

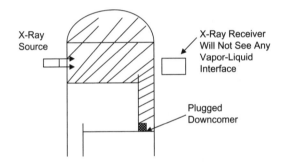

Figure 8-3   Examples of tray problems detectable by X-rays.

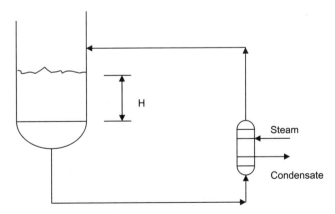

Figure 8-4   Thermosiphon reboiler operating principles. H, liquid height that causes reboiler to circulate. Feed to reboiler is all liquid. Vapor return line contains a mixture of liquid and vapor. Typically, this line contains 5 to 25% vapor. The amount of vapor in this line, which is also referred to as the percent of vaporization, is set by the pressure drop in the liquid and vapor return line.

There may be restrictions in the reboiler inlet or outlet piping. This will create additional frictional piping loss. The result of this problem will be similar to that discussed in the earlier paragraph, that is, the increased circuit pressure drop may result in a lower-than-design liquid level in the reboiler and/or a lower-than-design circulation rate.

There may be excessive elevation head in the tower bottoms. A high level in the tower bottom may cause cycles in a reboiler operation. This high level leads to a high circulation rate through the reboiler circuit. This will result in a high fluid density (low percentage vaporization) in the reboiler outlet. With the higher fluid density, the circulation rate will decrease. The reduced circulation rate will cause the percentage vaporization rate to increase, which will increase the circulation rate. The result of this can be a wildly cyclic operation of the reboiler circuit.

Another problem is that the tower level may be higher than the reboiler return line. While avoiding this seems obvious, the instrumentation is sometimes designed so that what seems to be a reasonable tower level is actually above the reboiler inlet.

If an inadequate reboiler performance is thought to be associated with a hydraulic problem, the best approach to generating a theoretically sound working hypothesis is to either redo or review the original hydraulic calculations. These calculations might be in error, or may be based on an incorrect length of equivalent piping. It is possible that the actual piping detail is radically different than that assumed by the process designer. This review should also include confirmation that the tower elevation relative to the reboiler is the same as it is in the original design. If the original calculations appear to be correct and consistent with the actual installation and operating details, then it is likely that there is a piping restriction somewhere in the reboiler piping circuit. The more common problem-solving tool of changing the level in the tower without first considering the hydraulic calculations will not be helpful. It is likely that these hydraulic calculations will require the assistance of a graduate engineer.

Another possible reboiler problem is associated with inadequate drainage of the steam condensate. This will cause a high level of condensate in the reboiler, resulting in some of the reboiler surface area being covered with condensate rather than condensing steam. This can also lead to poor stability in the reboiler control circuit. Some of the reasons for inadequate removal of condensate are described in the following paragraphs.

The steam traps can be improperly sized or can be malfunctioning. If this occurs, condensate will build up in the reboiler until the trap opens to discharge condensate.

Steam pressure modulation control systems often create problems when low pressure steam is used. If the reboiler is controlled by a control valve in the inlet steam line, it is possible for the pressure on the steam side of the reboiler to be lower than the pressure of the condensate return system. Since the steam pressure in the reboiler is lower than that in the condensate system, the condensate level in the reboiler will increase, covering some of the tube area with condensate. As the tube area is covered, the steam pressure must increase to compensate for this. As the steam pressure increases, the condensate is drained from the reboiler. This cycle will be repeated, leading to instability in the condensate removal system. This instability in the condensate removal system is often transmitted to the tower.

The temperature-driving force needs to be at an optimum level. The temperature difference between the heating medium and process side must not be too great or too small. A large difference (>100°F) can produce film boiling instead of nucleate boiling. This film boiling causes a vapor film to exist at the tube wall. Under some conditions, this will result in a much lower heat transfer coefficient than design. On the other hand, operating with a close approach between the heating fluid and process fluid can cause the reboiler to surge. That is, the reboiler will cycle between no heat input and greater-than-design heat input. This surging can be caused by depletion of volatile materials in the reboiler. If low pressure steam is used on a reboiler that heats a mixture of a volatile material (propane) and a nonvolatile material (octane), depletion of the volatile material will cause the boiling point of the process side to increase above that of the condensing temperature of the steam. This will cause the reboiler to stop condensing any steam until the inventory of the volatile material on the process side is replenished. In addition, the same effect can be observed due to depletion of water from a water low-volatility hydrocarbon two-phase mixture. In a similar fashion to that described above, this will result in a rapid change in boiling point and cause the reboiler to stop condensing steam until the water concentration is replenished.

Solving a reboiler problem can require almost no engineering, or it can require a great deal of engineering problem-solving skills. However, regardless of the technical complexity, the principles of problems solving can still be applied. The questions given in Chapter 6 will still be of value. In addition, simple instruments can be utilized for step 3 (develop a theoretically sound working hypothesis that explains the problem) and step 4 (provide a mechanism to test the hypothesis) of the problem-solving approach. Examples of these instruments are:

- An infrared thermometer, used to measure the temperature of a suspect steam trap.
- A gage glass and pressure gauge, used to measure the condensate level and pressure in a reboiler that is cycling excessively.

More complicated instrumentation might be used as follows:

- A highly sensitive pressure drop instrument might be connected to a high speed recorder or a process control computer with a rapid scan frequency. This could be used to diagnose reboiler hydraulic problems.

- A device such as a calorimeter that is capable of determining the enthalpy of a process stream could be used to measure the enthalpy of a stream leaving a reboiler. This will allow determination of the percentage vaporization that is occurring in the reboiler.

EXAMPLE PROBLEM 8-1

A propane/butane fractionation tower (T-1) had ceased to operate like it should. The data shown in Table 8-2 represents good versus current operation.

In Table 8-2, the term "saturated liquid" refers to a liquid stream that is at the boiling point of the liquid. For example, water at atmospheric pressure and 212°F is a saturated liquid. A schematic flow diagram is shown in Figure 8-5. In addition to the data shown above, the chronological information shown in the following paragraphs was available.

A routine repair and inspection downtime on the tower occurred sometime after the "good operation" data was recorded. During this downtime, the existing trays, which showed evidence of corrosion, were removed, and new trays with an identical design were installed. While this seemed to go well, the normal mechanical supervisor was off sick during the tray installation procedure. No obvious change in performance was noticed immediately. However, it was several weeks before full rates were achieved due to limited product demand.

The laboratory began using a different gas chromatograph (GC) than was used during the "good operation" period. This GC was thoroughly checked out in a series of cross checks with the old GC and other laboratory GCs. One of the laboratory technicians that had analyzed this sample for 10 years reported that he found a film of heavy hydrocarbons or oils in the GC after each analytical run on the distillate product. All flow meters and temperature instruments had been confirmed to be accurate.

Table 8-2 T-1 operating conditions

	Good Operation	Current Operation
Feed rate, lb/hr	100,000	100,000
Feed composition, wt % C_3	30	40
Feed enthalpy	Saturated liquid	Saturated liquid
Vapor boilup, lb/hr	79,500	124,000
Reflux rate, lb/hr	31,300	65,000
Reflux enthalpy	Saturated liquid	Saturated liquid
Distillate comp., wt % C_3	95	80
Control tray comp., wt % C_3	60	60
Bottoms, comp., wt % C_3	2	2

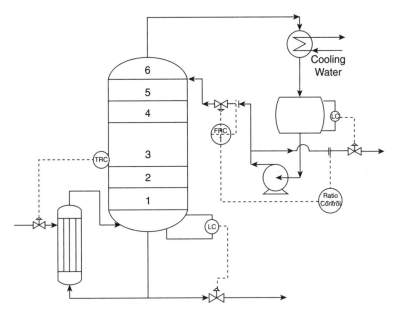

Figure 8-5 Schematic flow for Problem 8-1.

When the problem solver was assigned the problem, he began by using the five-step procedure discussed earlier. He combined steps 1 and 2 as follows:

Step 1: Verify that the problem actually occurred.

Since all indications were that there was a real problem, the problem solver simply combined the verification with the problem statement given in step 2.

Step 2: Write out an accurate statement of what problem you are trying to solve.

The problem statement that he developed was as follows:

Currently, T-1, the plant propane/butane splitter, is operating poorly based on historical standards. The current plant data indicates that the rectification section (the section of the tower above the feed tray) is not performing well even though the reflux rate is well above that required for good performance previously. There were several changes made recently. These changes consisted of the installation of new trays, the use of a new GC, and a feed composition change. All of these occurred prior to observation of the loss of fractionation. The loss of fractionation was not noticed until the product demand increased so that the tower began operating at design rates. Determine the reason for the poor performance of the rectification section of T-1.

Step 3: Develop a theoretically sound working hypothesis that explains as many specifications of the problem as possible.

When the chronological history was reviewed, it was tempting to believe that the trays were installed wrong. While this was a strong possibility, it did not represent a theoretically sound working hypothesis without adequate calculations and additional analysis. The questions in Chapter 6 were utilized, along with a tray stability diagram, to formulate hypotheses.

While this series of questions did not provide an exact diagnosis of the problem, it was apparent that the tower was operating well beyond the region of experience and operating directives. The first question that the problem solver sought to answer was "Do the current operating rates of the tower variables by themselves explain the poor tray performance?" Two types of calculations using the data were made to answer this question.

A tray stability diagram was developed and is shown in Figure 8-6. This diagram indicates that the top part of the tower is being overloaded and is probably operating in a flooded condition.

The internal tower vapor rate was calculated by material balance and heat balance using the principles discussed in Chapter 5. These calculated rates were then compared. If flooding was occurring, the vapor rate as determined by material balance would be higher than that determined by an overall heat

Table 8-3 Questions/comments for Problem 8-1

Question	Comment
Are all operating directives and procedures being followed?	No. Operating conditions appeared operating to be well outside the normal targets.
Are all instruments correct?	All instruments had been calibrated.
Are laboratory results correct?	The new GC had been thoroughly calibrated.
Were there any errors made in the original design?	Not applicable, since no design changes were made.
Were there changes in operating conditions?	Yes. In addition to the known changes, it is possible that whatever caused the percentage C_3 in the feed to increase might have also caused an extraneous component to be introduced into the tower.
Is fluid leakage occurring?	Damaged trays might cause internal tray leakage, which might explain the problem.
Has there been mechanical wear that might explain the problem?	No.
Is the reaction rate as anticipated?	Not applicable.
Are there adverse reactions occurring?	Not applicable.
Were there errors made in the construction?	Maybe; the tray installation might not have been correct.

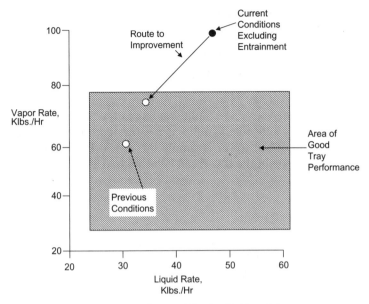

Figure 8-6   Tray stability for Problem 8-1.

balance. This is because as the tower floods, liquid would be carried over with vapor from the top tray into the reflux condenser. The total vapor (including entrained liquid) going into the condenser and accumulator would be greater than that estimated from a heat balance on the tower. Regardless of what control scheme was being used, this excess liquid must be pumped back to the tower and would show on the reflux flow meter. In summary, if the vapor rate calculated by material balance, and that calculated by heat balance, did not agree, it could be assumed that the excessive vapor rate based on the material balance must be related to liquid being carried out the top of the tower and showing up as reflux that is being pumped back to the tower.

The problem solver calculated the vapor and liquid loading in the top of the tower by material balance and heat balance as follows:

1. He calculated the distillate rate by material balance:

$$F = B + D \tag{8-1}$$

$$F \times X_F = B \times X_B + D \times X_D \tag{8-2}$$

where
F = feed rate, lb/hr
B = bottoms rate, lb/hr
D = distillate rate, lb/hr

X_F = feed concentration, wt %

X_B = bottoms concentration, wt %

X_D = distillate concentration, wt %

For the two cases given in Table 8-2, the calculated results are shown in Table 8-4.

2. He estimated the vapor and liquid loading in the top of the tower: Since the reflux is not subcooled (reflux is a saturated liquid), the liquid rate (L) in the top of the tower is simply the reflux. In addition, the vapor rate (V) in the top of the tower can be estimated as follows and is shown in Table 8-5.

$$V = L + D \tag{8-3}$$

3. Using the principle of "equal molal overflow," he calculated the top vapor rate (V) based on the heat input to the reboiler. The equal molal overflow principle states that for systems with minimal nonideality, the vapor rate throughout the tower expressed in mols/hour is equal as long as no other vapor or heat input is introduced. The tower feed is a saturated liquid (liquid at the boiling point), therefore there is no other vapor generated when the feed is added to the tower. In addition, there is no other heat input to the tower. Thus the equation below was used to estimate the vapor rate at the top of the tower from the heat input to the bottom of the tower. The results of this calculation for past and current operations are shown in Table 8-6.

Table 8-4 Estimated material balance rates

	Past Operation	Current Operation
Distillate (D), lb/hr	30,100	48,700
Bottoms (B), lb/hr	69,900	51,300

Table 8-5 Estimated vapor and liquid rates in top of tower

	Past Operation	Current Operation
Liquid (L), lb/hr	31,300	65,000
Vapor (V), lb/hr	61,400	113,700

Table 8-6 Top vapor rate based on heat input to bottom

	Past Operation	Current Operation
Vapor rate bottom, lb/hr	79,500	124,000
Molecular weight, top	44.54	46.23
Molecular weight, bottom	57.63	57.63
Vapor rate top, lb/hr	61,400	99,500

Table 8-7   Vapor rates calculated from material balances and heat balances

	Past Operation	Current Operation
Calculation method		
Top material balance, lb/hr	61,400	113,700
Bottom heat balance, lb/hr	61,400	99,500
Comments	Good check	Poor check

$$V_T = V_B \times M_T / M_B \qquad (8\text{-}4)$$

where

V_T = vapor rate at the top of the tower, lb/hr

V_B = vapor rate at the bottom of the tower, lb/hr

M_T = molecular weight of vapor top of the tower

M_B = molecular weight of vapor bottom of tower

The problem solver compared the top vapor rate as estimated from the overhead material balance and, from the vapor rate at the bottom, developed Table 8-7 as shown above.

The problem solver now had two independent calculations that indicated that the tower was flooding. The tray stability diagram indicated that the tower was operating in the flooding regime. In addition, the comparison of the heat balances and material balances indicated that there was more material leaving the top of the tower than could be accounted for by a heat balance. The fact that this approach had previously given a good comparison was proof that there was a significant change in the operation of the tower. The problem solver then developed the following hypothesis:

The poor performance of T-1 is due to flooding of the rectification section (top section of the tower) that is being caused by the excessive vapor and liquid loading. This excessive vapor and liquid loading might be caused by one of the following:

- *Operator error:* The reboiler steam rate or tower reflux rate was set too high by operator error.
- *Tray installation error:* There may have been an error in the tray installation which resulted in poor fractionation and caused the control system or operator to increase the tray loadings in an attempt to compensate for this. This caused the tower to operate in a flooded regime.
- *Foaming:* The presence of a surface-active material might have caused the tower to originally function poorly, which then caused a manual or automatic intervention to compensate for this uncovered problem. This intervention caused the tower to begin to operate in a flooded regime.
- *Other:* There may be other as yet undiscovered explanations for the changes which moved the tower operation into the flooding regime.

Note that in this problem hypothesis, the problem solver expressed the need to explore what may have caused the excessive tower loadings rather than to just assume that it was due to an operator or control system error. It should also be noted that, while the increased C_3 concentration in the feed to the tower is unlikely to be the root cause of the problem, it did indicate that the source of the C_3/C_4 liquid might have changed. This new source might have resulted in a different trace impurity or an increase in the concentration of an existing trace impurity that would cause foaming in the tower. The possible trace impurities in a C_3/C_4 liquid stream could include materials used for deicing exchangers (methanol), materials used for removing water (glycol) or materials used for neutralizing acidic compounds (amines).

Step 4: Provide a mechanism to test the hypothesis.

The problem solver decided to test the hypothesis by reducing the tray loadings to safe levels as indicated by the tray stability diagram. If this test returned the operation to "past operation" levels, then the mechanism to test the hypothesis becomes a partial solution.

He also considered what would happen if the operation did not return to the conditions and results previously experienced when the tray loadings were reduced. If that contingency occurred, he believed that additional tests or investigations would be required. These additional tests would include one or more of the following:

- Perform more detailed testing using some of techniques shown in Table 8-1.
- Conduct some sort of foaming test. This would probably require some type of pressure-rated laboratory equipment with a sight glass.
- Take X-rays of the appropriate parts of the tower to determine if tray damage is obvious.
- Determine if there has been a change in the source of the C_3/C_4 liquid and if there was any information on the trace impurities that might be present in this source.

If the reduction of the tray loadings does not solve the problem, it is likely that an expedient solution of reducing the feed rate to the tower may be required to ensure producing an overhead product that meets the specifications. However, this is not the solution to the problem, but only a stop-gap approach to allow the facility to continue to make on-specification product.

Step 5: Recommend remedial action to eliminate the problem without creating another problem.

The actual recommendation for remedial action will depend on the results of step 4. If reducing the tray loadings does not solve the problem, then the other tests listed in step 4 must be executed and carefully analyzed to determine the required remedial action.

If reducing the tray loadings does solve the problem, it should be recognized that the basic questions outlined in step 3 still remain. It is unlikely that improper tray installation would be a transient problem. It is also unlikely that an operator purposefully set the ratio controller outside the range of standard operations without a reason. Thus the most logical possibility is that there was a transient condition that was due to the presence of a trace impurity or some other external event. When the operator encountered this external transient event, he tried to respond to the unmeasured and unknown disturbance by increasing the ratio controller, which caused the tower to begin flooding. Since this transient condition might well recur, the tower should be monitored daily to determine when the problem reoccurs. It will be of value to be proactive and plan what analyses or actions will be taken when daily monitoring indicates that the transient condition is returning.

The test of reducing the reflux rate so that the operations were in the good tray-operating region was successful in returning the operations to normal rates and purities. A review of the possibility that the sources of the C_3/C_4 liquid had changed indicated that the source had changed, but since the problem only occurred as the production rate increased, it was impossible to connect this change with the change in tower operation.

As indicated earlier, in this circumstance it is necessary to monitor the tower closely and to develop a contingency plan to allow response to the likely return of the transient condition. The problem solver took the following actions:

- He developed a fractionation index which allowed him to follow the operation of the tower on an hourly basis.
- He developed a list of samples that were to be obtained the next time the fractionation index dropped below a specified value.
- He worked with the laboratory to develop new analytical techniques for the bomb samples to determine if there were surfactants present in any of these samples. The analytical techniques and GCs used for volatile hydrocarbons would likely not be sensitive enough for very low concentrations of surfactants.
- He developed a foaming test that would be used in the case of a significant decrease in the fractionation index.
- He collected the specified samples and used the new analytical techniques to determine the possible presence and concentrations of surfactants in the base case with good operations. He also tested the foaming potential of these samples.

Lessons Learned   While this problem is a fictitious example, it has many elements of real problem solving. One such example is that multiple events occur at the same time, making isolation of a single root cause difficult. The presence and subsequent disappearance of a trace impurity that leads to an unsuspected problem is also a potential real event. Therefore, there are lessons to be learned from this semi-fictitious example.

There was great value in doing calculations to prove that the tower was flooding. The tray loading diagram was a useful tool for problem solving as well as in selling the problem solution to management.

It should be recognized that, if daily monitoring had been utilized, the problem would have been spotted immediately, as opposed to having to wait until demand increased to the point at which the tower was required to operate at full capacity. While this may seem like an isolated occurrence, many industrial problems lie dormant until it is necessary to increase rates. Also note that it is never too late to begin a daily monitoring system.

This problem also illustrates that there will be occasions when the exact root cause of the problem cannot be determined. In these instances, the problem solver should develop a system that will be effective in collecting data when the next occurrence of the event happens.

While it could be argued that all problems have an obvious root cause, it should be recognized that multiple events that may seem to be related to the problem do occur. Thus the multiple events (the mechanical supervisor being sick, installation of a new GC, possible change in source of C_3/C_4 feed, and oil in the GC) that may be a cause of this problem are often typical of industrial problem solving. The problem solver must not discard any set of data or observations, but rather must incorporate them in his problem statement or problem analysis.

NOMENCLATURE

B Bottoms rate, lb/hr
D Distillate rate, lb/hr
F Feed rate, lb/hr
M_B Molecular weight of vapor bottom of tower
M_T Molecular weight of vapor top of the tower
V_T Vapor rate at the top of the tower, lb/hr
V_B Vapor rate at the bottom of the tower, lb/hr
X_F Feed concentration, wt %
X_B Bottoms concentration, wt %
X_D Distillate concentration, wt %

9

APPLICATION TO KINETICALLY LIMITED PROCESSES

9.1 INTRODUCTION

A kinetically limited process is a process that does not go to completion because it is limited by the speed of mass or heat transfer. Completion can be defined as equilibrium or a state that would be reached if an infinite amount of time were available. Examples of this process are heat exchange, reaction, and diffusion-limited operations. Diffusion-limited processes are those that do not reach equilibrium because the material of interest cannot easily flow through a membrane or from the inner part of a particle to the surface.

The purpose of this chapter is to illustrate how the five-step approach to problem solving can be used to solve problems in this type of process. The chapter shows how a generalized approach can be used to develop theoretically sound working hypotheses for any kinetically limited process. The emphasis is on utilizing this approach in an industrial setting. Thus the more theoretical approach of using multiple constants has been replaced with an empirical "lumped parameter" approach. A specific example of this that most process engineers and operators are familiar with is the use of an overall heat transfer coefficient rather than individual film coefficients. In theory, the overall heat transfer coefficient can be derived from individual film coefficients. These

Problem Solving for Process Operators and Specialists, First Edition. Joseph M. Bonem.
© 2011 John Wiley & Sons, Inc. Published 2011 by John Wiley & Sons Inc.

individual film coefficients are the coefficients for transferring heat from the fluid inside the tube to the tube wall, the coefficient for transferring heat across the thickness of the tube wall, and the coefficient for transferring heat from the outside tube wall to the process fluid flowing outside the tubes. The overall heat transfer coefficient can be calculated from these individual film coefficients. However, it can also be determined from vendor specification sheets or from plant test data when the exchanger is known to be clean. The overall coefficient determined in this fashion can be considered a "lumped parameter constant."

9.2 KINETICALLY LIMITED MODELS

Any kinetically limited process can be described by the generalized equation shown below:

$$R = C \times DF \qquad (9\text{-}1)$$

where

R = rate of change with time of the variable under study

C = a constant referred to as the "lumped parameter constant"

DF = driving force, or incentive for mass/heat transfer to occur

For a kinetically limited process, the lumped parameter constant C can be used for problem solving in multiple ways. The value of C can be determined on an hourly or daily basis and can be monitored as part of a daily monitoring system. Changes in C will be trigger points to start active problem solving. In addition, based on an estimated value of C, studies can be conducted to estimate the way changes in the driving force will impact the rate. This may allow process conditions to be modified to compensate for changes in C.

As indicated earlier, the heat transfer relationship shown below is the best known example of equation (9-1).

$$Q = U \times A \times \ln \Delta T \qquad (9\text{-}2)$$

where

$Q = R$ = rate of heat transfer

$U \times A = C$ = lumped parameter or, for heat transfer, the heat transfer coefficient multiplied by the area

$\ln \Delta T = DF$ = driving force or, for heat transfer, the log mean temperature difference

The log mean temperature difference ($\ln \Delta T$) is an engineering relationship between the temperatures in a heat exchanger. It can be calculated using equation (9-3), shown below:

$$\ln \Delta T = (\Delta t_1 - \Delta t_2)/\ln(\Delta t_1/\Delta t_2) \qquad (9\text{-}3)$$

where

$\Delta t_1 \doteq$ temperature difference between the hot fluid and cold fluid on the inlet side of the heat exchanger

$\Delta t_2 =$ temperature difference between the hot fluid and cold fluid on the outlet side of the heat exchanger

$\ln =$ natural log

The driving force can also be approximated by the difference between the average of the hot side and cold side temperatures.

A more complicated form of equation (9-1) is encountered for reaction and diffusion-limited drying. The driving force in these cases is often related to the concentration of a molecule or the difference between the concentration and equilibrium concentration. For example, for diffusion-limited drying, equation (9-1) becomes:

$$dX/dt = K \times (X - X_e) \qquad (9\text{-}4)$$

where

$dX/dt =$ rate of removal of solvent from a polymer

$K =$ lumped parameter constant which is somewhat related to diffusion (since it deals with mass transfer, it is also referenced in this chapter as a mass transfer coefficient)

$X - X_e =$ driving force

$X =$ actual concentration of solvent in a polymer

$X_e =$ concentration of solvent in the polymer in equilibrium with the vapor (at the same point in time as X)

Chemical reactions are generally "first order." This means that the reaction rate is proportional to the concentration of the reactants. For a simple reaction between two components where the reaction rate is first order, equation (9-1) becomes:

$$dX/dt = C_R \times X \times Y \qquad (9\text{-}5)$$

where

$dX/dt =$ rate of disappearance of component X

$C_R =$ lumped parameter constant (reaction rate constant)

$$X = \text{concentration of component } X$$
$$Y = \text{concentration of component } Y$$

Relationships for other kinetically limited processes can be developed starting with equation (9-1). The key to using this equation for developing theoretically sound working hypotheses is the correct selection of the driving force and use of the "lumped parameter" constant.

The need to select the correct driving force can be illustrated by equation (9-4). The correct driving force is the difference between the actual and equilibrium concentration. If the equilibrium concentration is ignored, the constant determined from plant operating data will be lower than the actual amount. Another example where the use of the incorrect driving force will give improper constants is estimation of heat transfer constants in a fuel-fired furnace. Heat transfer from the flames and refractory surfaces to the metal tubes (primarily radiant heat) depends on the driving force for radiant heat transfer rather than convective heat transfer. Whereas the driving force for convective heat transfer is simply the log temperature difference, the driving force for radiant heat transfer is expressed as shown below:

$$\text{DF}_R = T_G^4 - T_M^4 \tag{9-6}$$

where

DF_R = driving force for radiant heat transfer

T_G = absolute temperature of gas, °R

T_M = absolute temperature of tube metal, °R

Lumped constant parameters that are developed from plant data are unreliable when used in large extrapolations. A lumped parameter constant that can be developed from fundamentals will always have a broader range of applicability. These constants that are developed from fundamentals can be used regardless of how close the operating conditions approach those under which the constant was determined.

However, some constants can be determined from fundamentals only by such an elaborate procedure that it becomes impractical for plant problem-solving activities. For example, in diffusion-related drying (equation 9-4), the lumped parameter constant or mass transfer coefficient depends on the diffusion rate through the polymer, the effective length of the flow path through the polymer particle, the actual particle surface area, and the mass transfer coefficients from the polymer surface to the bulk of the gas. Most polymer particles are irregularly shaped and have a large number of internal pores. This makes determination of items such as particle surface area and flow path through the particle difficult, if not impossible, to determine. If only one of these individual variables is not available and cannot be developed or determined, then utilizing a lumped parameter empirical constant (overall mass transfer coefficient) provides the only means to evaluate diffusion-related drying. This overall mass transfer coefficient can be determined from plant data, from pilot plant tests, or from bench scale tests.

9.3 LIMITATIONS TO THE LUMPED PARAMETER APPROACH

The lumped parameter approach has sometimes been referred to as a "black box" approach. This implies that the person using this approach does not understand the details of what is occurring inside the "black box." For example, the use of an overall heat transfer coefficient does not take into account the individual film side coefficients, metal resistance, and fouling factors. However, in most process plant problem-solving activities, the significant observation is the change in the overall heat transfer coefficient. Whether the decrease of heat transfer coefficient is due to the tube- or shell-side film coefficient decrease can usually be determined based on experience. For example, the decrease in overall heat transfer coefficient for a fractionating tower overhead condenser that utilizes cooling tower water is almost always due to fouling on the cooling water side. Similar logic would apply to the use of the lumped parameter approach for problem solving with other kinetically limited processes.

The limitation of this approach is more severe when designing new facilities based on a lumped parameter coefficient determined from an operating plant. If a more fundamental analysis is not done, the lumped parameter approach can lead to design errors. An example of this is the design of a specialized vertical condenser for a new plant, based on a condenser in identical service in an existing plant. The condensation in this vertical condenser took place on the shell side and cooling water was utilized in the tubes. The new plant had a slightly higher capacity. To evaluate the overall heat transfer coefficient from pure fundamentals would have required an elaborate model that involved condensate thickness at various points in the vertical exchanger as well as the traditional resistances such as water side coefficient, condensing side coefficient, and metal resistance. Rather than do this highly theoretical analysis, however, the design team chose to utilize the overall heat transfer coefficient experienced in the existing plant. Based on the desire to minimize the overall height of the structure in the new plant, the height to diameter (H/D) ratio of the vertical condenser was reduced. In order to compensate for this reduced H/D ratio, the tube diameter and exchanger shell diameter were increased. No consideration was given to the effect of these increased diameters on the overall heat transfer coefficient. When the unit was started up, however, it was noticed that the heat transfer coefficient was 20% below anticipated values. This underscores the need for geometric similarity when basing a design on empirically developed constants.

In spite of the limitations, when considering plant-related problem solving, the utilization of an empirically derived lumped parameter constant along with a theoretically correct driving force is almost always adequate, for the three reasons given below:

1. Daily monitoring of the process requires only an empirical constant.
2. Time is always critical. Laboratory or elaborate investigations to develop fundamental data are rarely appropriate.
3. If it is desirable to make operating changes, the new conditions are generally not extraordinarily different from current conditions. Thus there is minimal danger of extrapolation.

9.4 GUIDELINES FOR UTILIZATION OF THIS APPROACH FOR PLANT PROBLEM SOLVING

The "lumped parameter constant" approach for monitoring and solving plant problems is a powerful tool. However, like all tools, it must be used in an appropriate fashion. Some guidelines for using this tool are as follows:

1. Develop a meaningful driving force that is theoretically correct. It should be noted that there is often a difference between a driving force that appears logical and one that is theoretically correct. Logic cannot be substituted for the utilization of sound engineering fundamental knowledge. Without a meaningful driving force, the empirically developed lumped parameter constant will not be valid over any range of data.
2. Monitor this lumped parameter constant on a daily basis and over as wide a variety of conditions as possible. If the constant varies, look for correlations between the constant and independent variables that make theoretical sense. For example, a correlation between the lumped parameter constant in a diffusion-limited drying process and gas rates makes theoretical sense. The increased gas rate should increase the rate of mass transfer. If a lumped parameter constant varies with no process changes, it is an indication that an extraneous factor has caused a process deviation. If this extraneous factor can be eliminated, then the constant should return to normal. If this factor cannot be discovered or eliminated, then process changes will be required.
3. If it appears necessary to recommend a change in the driving force to increase the kinetically limited rate, minimize the length of the extrapolation that uses an empirically developed lumped parameter constant. That is, increasing a variable by 20–50% will likely be acceptable. But increasing the variable by 100% may make the approach based on the empirically developed lumped parameter constant invalid.

EXAMPLE PROBLEM 9-1

A polymer plant was experiencing problems in stripping the residual solvent from the polymer product. A counter flow agitated dryer with pure nitrogen sweep gas was utilized to strip the solvent. All operating conditions appeared to be normal. In addition, all instruments were checked and appeared to be accurate. A timeline indicated that the problem seemed have begun when a new catalyst was introduced into the process for a plant test. Even though the utilization of the new catalyst was considered a test, it was mandatory that the plant be switched to this new catalyst as soon as possible. One of the key advantages of the new catalyst was the higher bulk density product that could be produced with the catalyst. This had been well demonstrated in pilot plant studies. No drying studies had been conducted in the pilot plant. Drying capability was thought to be associated with sweep gas rate, residence time, and temperatures, all of which were not changed when the plant was switched to the new catalyst.

The problem solver was asked to determine what could be done to reduce the solvent levels to the previous concentrations while continuing to operate with the new catalyst. Operating conditions were as in Table 9-1.

In addition to the plant data shown above, laboratory results indicated that the equilibrium relationships for polymers produced with the old and new catalysts were identical and could be expressed as shown below:

$$X_E = 295,000 \times (Y \times \pi/VP)^{1.5} \tag{9-7}$$

where

$\quad\quad X_E$ = equilibrium concentration of solvent in polymer

$\quad\quad Y$ = concentration of the solvent in the vapor phase

$\quad\quad \Pi$ = total pressure on the system

$\quad\quad VP$ = vapor pressure of the solvent at the dryer temperature

Table 9-1   Operating conditions with old and new catalyst

	Old Catalyst	New Catalyst
Polymer rate, lb/hr	20,000	20,000
Polymer density, lb/ft³	25	28
Pure nitrogen sweep gas rate, lb/hr	1000	1000
Dryer temperature (isothermal), °F	200	200
Hexane vapor pressure, psia	28.7	28.7
Dryer pressure, psia	15	15
Inlet hexane content, ppm	1000	1000
Outlet hexane content, ppm	25	75

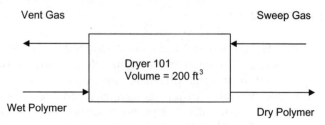

Figure 9-1 Schematic sketch of dryer.

A schematic sketch of the dryer is shown in Figure 9-1.

The problem solver began working on the problem using the five-step problem-solving approach. There was minimal need to verify that the problem had actually occurred. However, he did not skip this step.

Step 1: Verify that the problem actually occurred.

The problem solver confirmed that all of the meters were correct and that the hexane content was indeed three times the normal levels.

Step 2: Write out an accurate statement of what problem you are trying to solve.

He wrote out the following problem statement.

> The polymer leaving Dryer 101 contains excessive amounts of solvent. This condition started soon after the introduction of a new catalyst to the polymerization reactors. All operating conditions are normal and the instrumentation is correct. It is desirable to use the catalyst throughout the plant as soon as possible. Determine what operating condition changes or what new equipment is required to achieve 25 ppm hexane in the final product at the normal polymer production rate.

Step 3: Develop a theoretically sound working hypothesis that explains as many specifications of the problem as possible.

Referring back to the list of questions given in Chapter 6 was helpful in formulating possible hypotheses. The problem solver reviewed these questions and formulated appropriate comments for this example problem, as shown in Table 9-2.

As might be expected, these questions pointed strongly to the idea that the problem began when the new catalyst was introduced to the process. They also provided some possible explanation for why the new catalyst would cause a

Table 9-2 Questions/comments for Problem 9-1

Question	Comment
Are all operating directives and procedures being followed?	All appeared to be correct and being followed.
Are all instruments correct?	The instruments had allegedly been calibrated.
Are laboratory results correct?	Yes.
Were there any errors made in the original design?	Not applicable.
Were there changes in operating conditions?	New catalyst was started.
Is fluid leakage occurring?	Not applicable.
Has there been mechanical wear that would explain problem?	Not applicable.
Is the reaction rate as anticipated?	The new catalyst might cause the polymer particles to be less porous. This would cause a lower than anticipated "mass transfer coefficient."
Are there adverse reactions occurring?	See above.
Were there errors made in the construction?	Not applicable.

drying problem to occur. It was theorized that the new catalyst created a less porous particle. The lower porosity would slow the diffusion of the solvent through the particle. This less porous particle would also have a higher particle density, which would explain the higher bulk density. It should be noted that if this list of questions were not utilized, the change in catalyst would have been pinpointed as the cause, but the explanation of the formation of a less porous polymer particle might have been missed.

As indicated in equation (9-4), the relationship between rate of drying and driving force for a diffusion-limited drying process can be expressed as follows:

$$dX/dt = K \times (X - X_e) \qquad (9-4)$$

Since the problem statement indicates that all operating conditions were normal and laboratory results indicated that the equilibrium relationship for both catalysts was the same, there must be a difference in the K values between the two catalysts.

While a working hypothesis that the K value has changed is a perfectly valid hypothesis, it does not help determine what the next step should be because it does not specify the magnitude of the change. If the magnitude of the change is not known, it will be impossible to determine if operational changes or addition of new equipment will be required to allow operation with the new catalyst. Thus the theoretically sound working hypothesis must include an estimate of the magnitude of the change in the mass transfer coefficient (K value).

While it is beyond the scope of this book to discuss integral calculus, the value of K can be estimated by simple integration using the rules of integral calculus, if X_e (the equilibrium concentration of solvent in the polymer) is equal to zero. In this special case, integration yields the following:

$$X_f/X_o = e^{-Kt} \tag{9-8}$$

where

X_f = concentration of solvent in the outlet polymer
X_o = concentration of solvent in the inlet polymer
K = rate constant or mass transfer coefficient, 1/min
t = amount of time in the dryer, min

For the general case where X_e is not equal to zero, numerical integration must be used to determine the unknown variable. Numerical integration is a mathematical technique that involves calculating a variable over a small segment of a fixed size piece of equipment. The values for each of the small segments can then be added to give the final answer.

This numerical integration can be used to determine the mass transfer coefficient (K) or the outlet concentration of solvent in the polymer (X_f). This numerical integration can be developed by visualizing a dryer segment as shown in Figure 9-2. The equilibrium concentration in Figure 9-2 can be determined by a Henry's Law-type relationship, as shown below:

$$X_e = C_E \times Y \tag{9-9}$$

where

C_E = a constant determined from experimental data or application of theoretical principles
Y = vapor phase composition at any point in the dryer

For the example problem, the equilibrium relationship shown in equation (9-7) can be utilized in place of that shown in equation (9-9) to determine the equilibrium concentration based on the vapor phase concentration.

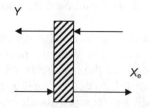

Figure 9-2   Dryer calculation segment.

In order to use numerical integration, equation (9-4) must be transformed as shown below. This will be the technique used when X_e is not equal to zero.

$$dX/dt = K \times (X - X_e) \qquad (9\text{-}4)$$

$$dX/(X - X_e) = K \times dt \qquad (9\text{-}10)$$

By integral calculus:

$$\int dX/(X - X_e) = Kt \qquad (9\text{-}11)$$

For this problem, it is desired to obtain the K value for each catalyst system. The term $\int dX/(X - X_e)$ must be developed by splitting the dryer into several small sections as described in Figure 9-2 and determining the value of $dX/(X - X_e)$ for each segment. The value of the left hand side of equation (9-11) will then be the sum of the values for each segment. To determine the K value from this relationship using numerical integration requires the following steps using the data given above and a spread sheet:

1. Use values of the polymer and gas rates along with the inlet and outlet hexane concentrations in the polymer to determine the concentration of hexane in the outlet gas. Note that the data indicated that pure nitrogen was used as the sweep gas. The material balance concepts discussed in Chapter 5 will be used in this step.

2. Calculate the total residence time in the dryer assuming no "back-mixing" takes place. The assumption of no back mixing is equivalent to assuming that each polymer particle has the exact same residence time in the dryer. While this is highly idealistic, the K value will take the presence of back mixing into account. The residence time can be calculated by dividing the polymer holdup in the dryer by the polymer rate. The polymer holdup in the dryer is often specified by the dryer vendor or can be calculated if the polymer level in the dryer is known.

3. Split the dryer into several segments (100 to 500) and calculate the dX for each segment. The change in the volatiles concentration (dX) will simply be the total change in volatiles concentration divided by the number of dryer segments. The time increment (dt) can be determined in a similar fashion.

4. From the relationship given in equation (9-7) and the outlet vapor composition, calculate the X_e of the polymer leaving the first segment.

5. Since this segment represents only an exceptionally small part of the dryer, the actual solvent concentration (X) can be considered constant throughout this segment. For this small segment:

$$(Kdt)_i = dX/(X - X_e) \qquad (9\text{-}12)$$

In equation (9-12), the term $(Kdt)_i$ is simply the product of the lumped parameter drying constant (mass transfer coefficient) and the residence time in the small segment. The lumped parameter drying constant is assumed to be the same in all segments.

6. Using dx, the gas and polymer rates, and the gas concentration leaving the first segment, calculate the gas concentration entering the segment by a material balance around the segment. This will be the gas concentration leaving the next segment.

7. Continue this process until the same calculations have been made for each segment. Then K can be calculated as follows:

$$K = \xi\,(Kt)_i\,/t \qquad\qquad (9\text{-}13)$$

That is, the overall K can be determined by summing up the Kt term from each segment as calculated from equation (9-12) and dividing the sum by the total residence time.

When the problem solver performed these calculations for the two cases described earlier, the following values of K were obtained as shown in Table 9-3. Extensive data was available from computer archives so that the value of K over the period of interest could be evaluated. When this was done, Figure 9-3 clearly showed that the mass transfer coefficient decreased when the new catalyst was introduced into the system.

Table 9-3   Results of numerical integration

Catalyst	K value, 1/min
Old	0.25
New	0.16

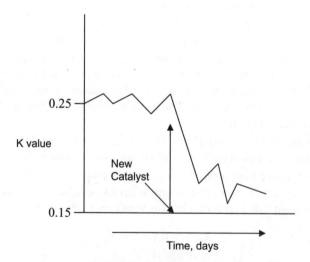

Figure 9-3 K value vs. time.

A theoretically sound working hypothesis that fits the data would be:

The recent loss of drying capability is associated with a decrease in the mass transfer coefficient (K) that occurred when the change to the new catalyst occurred. It is believed that this change is due to the lower porosity of the polymer particle produced with this catalyst. Evidence that supports this is the 35% lower K value and the higher bulk density.

Step 4: Provide a mechanism to test the hypothesis.

At this point, the problem solver gave consideration to the question of "optimum technical depth." Several alternatives were available to test the hypothesis. The optimum alternative depended on the cost of the solution, the confidence level required and the cost of continuing poor drying performance. In this polymer process, the customer specifications and/or needs were considered more important than operating at full capacity. Thus, in this case, the cost of the poor drying performance was the lost profits associated with operating at lower than design production rates. Alternatives for testing the hypothesis that were considered were:

- Conducting a test run using the old catalyst to confirm that the old catalyst really had a K value higher than the new catalyst.
- Using the K value for the new catalyst to estimate what changes in operating conditions would be required to reduce the hexane concentration from 75 ppm to 25 ppm.
- Modifying the new catalyst so that it produced a polymer similar to that produced by the old catalyst.

In order to determine if operating conditions could be modified to reduce the amount of hexane to 25 ppm, the problem solver made calculations using the approach described earlier. However, in these calculations, the K value was known to be 0.16 1/min. The value to be estimated was the outlet solvent concentration. The same spread sheet and approach was used. The problem solver changed the operating variables as desired and then varied the outlet solvent concentration in the polymer until the calculations based on the spread sheet indicated a K value of 0.16 1/min. Based on the calculations, he concluded that increasing the sweep gas by a factor of 10 (1000 lb/hr to 10,000 lb/hr) would reduce the hexane content from 75 to 65 ppm. This was obviously not a valid approach. The calculations also indicated that if the residence time in the dryer were increased from 17 min to 24 min, the hexane concentration would be reduced to 25 ppm. This could be accomplished in a plant test by reducing the polymer production rate to 14,000 lb/hr. The volumetric holdup in the dryer was constant, so the only way to increase residence time was to reduce the production rate. The calculations also indicated that the approach of increasing the dryer temperature was not a fruitful route to improved performance. They showed that if the dryer temperature were increased from 200°F to 225°F, at the same residence time and sweep gas rate, the hexane concentration would decrease from 75 to 65 ppm.

From these calculations, it was obvious that the only valid test of plant operating conditions was to reduce the polymer production rate from 20,000 to 14,000 lb/hr. The plant test of operating at reduced rates was the preferred test compared to a plant test of using the old catalyst, since it would be quicker. Introducing the old catalyst into the reactors would require significant residence time to completely displace the new catalyst from the entire system prior to the dryer. In addition, it was mandatory to continue to use the new catalyst. The plant test of returning to the old catalyst just confirms that the new catalyst is responsible for the problem. A plant test of increasing the dryer residence time by reducing the production to 14,000 lb/hr was successful at reducing the hexane content from 75 ppm to 25 ppm. The K value did not change as the residence time was increased.

Step 5: Recommend remedial action to eliminate the problem without creating another problem.

Since the plant test of operating at reduced rates was successful and a study of changing the catalyst to produce a more porous structure indicated that other desirable catalyst attributes would be lost if the porosity was changed, it was decided to add an additional dryer. This additional dryer would provide the necessary residence time. Since the dryer addition would take several months, it was decided to return to using the old catalyst in the interim. When use of the old catalyst resumed, the mass transfer coefficient increased from 0.16 1/min to the previous value of 0.25 1/min.

Lessons Learned If daily monitoring of the K value had been done, it would have been possible to determine immediately that there was likely a drying problem associated with the new catalyst. As often happens in a plant test, there is only a minimal potential problem analysis conducted prior to the test. This often leads to a panicky approach to problem solving. Chapter 12 discusses approaches for conducting successful plant tests.

If drying problems had been anticipated prior to the plant test of the new catalyst, it would have been possible to use laboratory techniques to determine the magnitude of the difference in mass transfer coefficients. An apparatus such as a thermal gravimetric analyzer (TGA) could have been used to determine the mass transfer coefficient for polymer produced with both catalyst systems. While the exact absolute values of the mass transfer coefficient determined in this fashion may not have been the same as those in the plant, the relative values would have been accurate. That is, the TGA would have predicted that the mass transfer coefficient (K) would have been 35% lower with the new catalyst.

The model of the dryer that was developed is a good example of a simple but valuable tool for doing process analysis work. The driving force of the actual concentration less the equilibrium concentration $(X - X_e)$ is theoretically correct and easy to determine. Experimental relationships such as equation (9-7) may not be available. However, there are calculation methods available to approximate the relationship between the vapor phase concentration of the solvent and the concentration of solvent in the solid. One of these techniques is discussed in Chapter 13. The mass transfer coefficient can be determined from plant data. It is subject to error if there is a significant change in the polymer morphology. For example, when the catalyst was changed, the mass transfer coefficient changed. However, it did not change when the production rate was reduced.

The calculations that were done to estimate the impact of changing plant operating conditions were performed very quickly once the spreadsheet described above was completed. The more expensive alternative of running multiple plant trials to test any reasonable hypothesis would have taken much more time and been much more expensive.

EXAMPLE PROBLEM 9-2

Several of the tube metal temperature indicators in an operating furnace were indicating considerably higher temperatures than anticipated when compared to historical values. Operations personnel were not concerned since this furnace was operating at normal heat duties and normal process fluid temperatures, and the furnace had never shown any tendency to foul. Fouling often occurs inside the tubes of an operating furnace as the high temperature of the tubes causes the material in the tubes to decompose and deposit a thin layer of fouling material on the tube surface. This reduces the heat transfer coefficient and results in the tube temperatures becoming even hotter. The operators concluded that the tube metal temperature indicators must have failed due to the harsh furnace environment. The problem solver did not believe that all of the tube metal indicators could have failed. A schematic diagram of this problem is shown in Figure 9-4.

The current operating data and the historical operating data are shown in Table 9-4. There was no other data available in the computer archives. Even though this problem was discovered by the problem solver rather than being delivered to him by a request from operations management, it is still a valid problem that can be approached using the five-step procedure. Because of the belief of operations personnel that there was not really any problem, step 1 had to be approached in a different fashion than it has been in problems described previously.

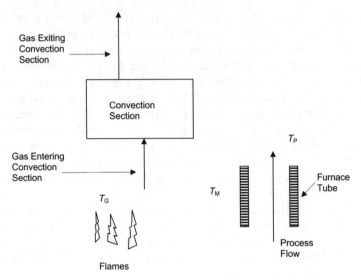

Figure 9-4 Furnace schematic. T_G, gas temperature; T_M, tube metal temperature; T_P, process temperature.

Table 9-4 Furnace data

	Historical	Current
Tube metal temperature, °F	600 to 650	750 to 900
Gas temperatures		
Entering convection section, °F	1200	1300
Leaving convection section, °F	650	700
Excess oxygen, %	15	15
Heat absorbed, MBTU/hr	80	80
Circulation to furnace k-lb/hr	700	700

Step 1: Verify that the problem actually occurred.

Step 1 required that careful examination be given to operation management's contention that there was really no problem with the indicated high tube metal temperatures. It was important that the problem solver convince himself as well as the operations personnel that there was a problem. If operations personnel were going to be expected to cooperate on any plant tests, they needed to be convinced that there was a problem. One of the first things that the problem solver had to do was understand how the tube metal temperature indicators were designed and installed. These thermocouples are different from normal ones in that they must measure the temperature of the furnace tube without being impacted by the radiant heat from the furnace flames. The problem solver obtained information from both instrument technicians and instrument engineers to make sure that he understood the technology of tube metal temperature indicators. Once he understood this technology, he recognized that while tube metal temperature indicators do fail, they almost always fail in such a fashion that the failure is obvious. That is, they read ridiculously high or low temperatures. Failed tube metal temperature indicators will normally have "up scale burnout," that is, they will read ridiculously high. He also recognized that if fouling had occurred on the inside of the tube, there would be other indications of fouling besides elevated tube metal temperatures. The problem solver decided to investigate whether there were other indications of fouling. He used the basic equation shown below to describe the radiant heat transfer that is occurring in the furnace:

$$Q = C_1 \times (T_G^4 - T_M^4) = C_2 \times (T_M - T_P) \tag{9-14}$$

where

Q = heat transferred to the process (this is equal to the heat transferred from the gas flame by radiation. In addition, it must also be equal to the heat transferred through the tube to the process fluid)

T_G = absolute temperature of the gas in the radiant section of the furnace

T_M = absolute temperature of the tube metal

T_P = absolute temperature of the process fluid in the tubes

C_1 = a constant that is related to the type of flame (should not change unless the type of fuel changes)

C_2 = product of the heat transfer coefficient and the area of the tubes (will only change if the heat transfer coefficientchanges)

In equation (9-14), two different kinds of heat transfer are represented. Radiant heat transfer is the type of heat that is transferred through any type of fluid. It does not depend on the two bodies touching each other (conductive heat transfer) or on flow patterns (convective heat transfer). It is proportional to the difference in the absolute temperature of the hot source and the cold receiver, each raised to the fourth power. The warming impact of sunlight is an example of radiant heat transfer. Convection and conduction depend on the difference in temperature in either absolute or conventional units. As shown in this equation, the radiant heat transferred from the flame to the tube must equal the convective heat transferred from the tube to the process fluid.

From equation (9-14), the problem solver noted that if the tube metal temperature (T_M) increases at constant heat duty (Q), it must be due to a decrease in the convective heat transfer coefficient, since the area of the tubes will not change. This decrease in heat transfer coefficient could be caused by a change in fluid circulation rate or tube fouling. Since the process flow rate to the furnace is constant, the decrease in heat transfer coefficient is likely due to tube fouling. In order to achieve the same heat duty at the higher tube metal temperature, the gas temperature (T_G) must also increase so that the radiant heat transferred equals the convective heat transferred. Thus the increased gas temperature shown in Table 9-4 was confirmation that the tube metal temperature had increased. A secondary confirmation was the higher-than-normal gas temperature leaving the convection section, also shown in Table 9-4. The higher-temperature gas entering the convection from the radiant section would not be cooled to as low a temperature as historical data would indicate since the heat transfer area in the convection section was fixed. Using this logic, the problem solver convinced the operations management that there might be a fouling problem in the furnace tubes.

Step 2: Write out an accurate statement of what problem you are trying to solve.

An accurate statement of the furnace problem is as follows:

> Tube metal temperatures on the process furnace indicate that the tubes are partially fouled with deposits. Since a furnace shutdown to inspect and possibly replace the tubes will require a plant shutdown, a detailed investigation was undertaken to confirm that the tube metal temperatures are higher than normal. This investigation indicated that the tubes are very likely fouled. There is no archived data to indicate when the fouling began or if it was a gradual fouling or a one-time event. While there is strong evidence that the furnace tubes are partially fouled, it is desirable to provide operations management with further evidence that the furnace is fouling and determine what steps should be taken to avoid a tube failure with a subsequent furnace fire. In addition, it will be necessary to determine why the tubes in this furnace are fouling when, based on operation personnel's memory, this has never happened in the past.

Step 3: Develop a theoretically sound working hypothesis that explains as many specifications of the problem as possible.

In this example problem, step 3 is different than it has been in previous problems. The gist is to convince operations that a problem exists with the furnace. As discussed earlier, the logic path that was developed was partially successful in convincing operations management that there really was a fouling problem. In addition to this, consideration was given to alternative means to measure the tube metal temperatures. One possibility was to use infra-red temperature measurements through a furnace peephole. Besides confirming the tube metal temperature measurements, this would also allow determination of where the hottest points were. The measurements based on tube metal temperature indicators are only single points in a few tubes; a much hotter point might exist in another location. An infra-red scan was conducted on the furnace tubes. The thermocouples were verified and even higher temperatures were discovered at spots where no tube metal indicators existed.

The furnace history was considered by asking such questions as:

- When did the fouling begin?
- Did it coincide with an upset of a magnitude never before experienced?

Table 9-5   Questions/comments for Problem 9-2

Question	Comment
Are all operating directives and procedures being followed?	All appeared to be correct and being followed.
Are all instruments correct?	The instruments had allegedly been calibrated.
Are laboratory results correct?	Not applicable.
Were there any errors made in the original design?	Not applicable.
Were there changes in operating conditions?	No.
Is fluid leakage occurring?	Not applicable.
Has there been mechanical wear that would explain problem?	Fouling of tubes.
Is the reaction rate as anticipated?	Not applicable .
Are there adverse reactions occurring?	Fouling is caused by reaction to produce coke. This reaction might be catalyzed by material present in the process. The fouling rate seems much higher than it was in previous experience.
Were there errors made in the construction?	Not applicable. It had been 36 months since mechanical work was done on furnace.

- Has the furnace been operating for a longer continuous period than it was in previous experience?

In addition to these questions, the questions from Chapter 6 were used to help develop a working hypothesis.

The following was developed as a working hypothesis:

The higher-than-normal tube metal temperatures in the furnace appear to be correct based on both gas temperatures and an examination of the instrument specifications which indicates that the thermocouples are 'up-scale burnout' (if they fail their readings will be ridiculously high). In addition, an infra-red scan indicates that these thermocouple measurements are correct and that there are even higher temperatures on other tubes. A review of the furnace history indicates that the current run (36 months without water washing the tubes) is nearly twice the length of previous periods of operation. Historically, the furnace tubes were water washed at every 18 month furnace inspection downtime. However, they were not water washed at the last downtime because it was expedient to get the furnace back in service. It is believed that this water washing removes the fouling material before it becomes thick enough to impact the furnace operation. The fouling is believed to be initiated by process catalyst residue that is entrained during process upsets.

Step 4: Provide a mechanism to test the hypothesis.

In this case, because of the risk of a tube failure, the only mechanism was to shutdown the furnace and water wash and inspect the tubes. The downtime was timed to coincide with a period of reduced product demand so that the lost revenue was minimized. An inspection of the tubes indicated that many were approaching the point of incipient failure. If the tube metal temperatures had been ignored (as proposed by operating personnel), it is highly likely that a furnace fire would have occurred. This not only would have caused serious damage to the furnace, but is likely to have occurred at a time when full production rate was required.

Step 5: Recommend remedial action to eliminate the problem without creating another problem.

Three remedial actions were recommended. The first was to carefully monitor the tube metal temperatures during the interim period between the discovery of the problem and the furnace shutdown. This allowed for a controlled plant shutdown to replace furnace tubes rather than an emergency shutdown during a period of high product demand. The second was to water wash the furnace tubes every 18 months. The third recommendation was to monitor the C_2 value as determined from equation (9-14) on a daily basis. This would provide information on how the fouling occurred. It could have been a one-time event that caused a sudden decrease in the value, or a slow continuous decrease.

Lessons Learned   This problem illustrates the value of a careful analysis of all operating data as opposed to only working on problems that operations or mechanical personnel consider to be important. The initial reaction of operations personnel to ignore the high tube metal temperature indicators was carefully considered as part of step 1. In addition, the actions of the problem solver to understand the technology of tube metal thermocouples was consistent with the principle of knowing the technology before trying to solve problems.

This problem also illustrates the fact that historical data is of great value in solving problems. However, conclusions (e.g., the furnace tubes never foul) based on historical data or memories that are not well supported are counterproductive. In this case, all of the current data indicated that the furnace tubes were indeed fouled. Thus it was apparent that the conclusion based on memories and/or historical data that indicated the tubes could not be fouled was in error. It is also likely that the tubes had been fouling in the past, but because the C_2 value was not being monitored, the fouling went undetected. The fouling material was likely removed every 18 months when the tubes were water washed.

If the problem solver had not been aggressive to the point of creating tension, it is likely that a furnace tube failure with a subsequent furnace fire would have occurred.

EXAMPLE PROBLEM 9-3

A plant that produced and shipped a polyolefin polymer in railroad hopper cars was contacted by a TV station with a video of one of their hopper cars on fire. These hopper cars were loaded at the manufacturing plant by discharging pellet storage bins directly into the closed hopper cars. While the hopper cars were closed to protect the product from the elements (rain, sunshine, wind, debris, etc.) there were vents in the front and back of the each car. Thus the vapor space in the hopper car was essentially air at the ambient conditions plus any hydrocarbon that evolved from the polymer.

When the company public relations contact appeared on the TV station to discuss the hopper car fire, he indicated that the particular polymer was not flammable and that could not have possibly caught on fire without some external source such as sabotage or excessive heat generated by a mechanical failure or malfunctioning of the hopper car equipment.

Unfortunately, shortly after this occasion, several other hopper cars arrived at different customers with blackened vapor vents and some charred polymer, indicating that there had been a fire of a limited magnitude during transportation. The material that was being shipped in the hopper cars was in the form of small pellets that had been extruded after being produced in the polymerization section. The polymerization section stripped the solvent and unreacted monomer from the polymer at a temperature of 220°F and a dryer residence time of 30 min. In addition, during the extrusion operation, the polymer was heated in the extruder to 550°F before being pelletized and cooled in a water bath. It seemed very unlikely that there could be sufficient residual monomer or solvent that would create a fire. In addition, operations personnel knew that when the hopper cars were loaded, they could see a great deal of static electricity being discharged inside the hopper car. They reasoned that if there was an explosive atmosphere inside the hopper car, it would be ignited during the loading operation. They believed that the maximum concentration of hydrocarbons in the vapor space would occur during loading, and during transit the vents would create a sweep of air through the vapor space, reducing the concentration of hydrocarbons.

However, because of the multiple indications of fires in the hopper cars transporting the polymer, the operations personnel requested that a technical evaluation be made of what was causing the hopper car fires.

The problem solver used the five-step procedure to methodically develop a problem solution as described in the following paragraphs.

Step 1: Verify that the problem actually occurred.

There was no doubt that something unusual had happened. The problem solver decided that he needed to know what product had been loaded into the hopper cars that experienced the major fire and evidence of minor fires. When he investigated the loading records, he found that all of the unusual incidents had occurred when a specific product (experimental product X3) was being produced and loaded into the hopper cars. However, not all hopper cars loaded with this product showed evidence of flash fires. He recognized that this product was one that had to be produced at reduced rates to ensure that the volatiles (unreacted monomer and solvent) were adequately stripped. He knew that rates had recently been increased by modifications to the operating procedures and directives.

Step 2: Write out an accurate statement of what problem you are trying to solve.

The problem solver wrote out the following problem statement:

> There are indications that flash fires are occurring in some of the railroad hopper cars loaded with experimental product X3. The evidence of these flash fires was noticed after the recently instituted operating procedures and directives were put into use. While the actual damage was minimal, there was a great deal of customer dissatisfaction. Determine the cause of the flash fires. This analysis should include the observation that not all hopper cars loaded with this product showed signs of a fire. Once the cause has been determined, provide recommendations for eliminating the flash fires. In addition, any recommendations should provide for shipping a product that meets Department of Transportation (DOT) regulations.

Step 3: Develop a theoretically sound working hypothesis that explains as many specifications of the problem as possible.

In addition to the observations described above, the questions from Chapter 6 were used to help develop a working hypothesis. These questions are shown in Table 9-6.

After reviewing the data and the questions, three possible hypotheses were developed as follows:

1. The new operating directives and procedures were causing excess quantities of unreacted monomer or solvent to be left in the polymer, which then accumulated in the hopper car and formed an explosive mixture. This mixture could then be ignited by an undefined ignition source.

2. There were changes in the catalyst which now resulted in a less porous polymer particle. Thus the temperature and residence time that were previously used successfully in a plant test to demonstrate the new procedures for X3 were no longer adequate. Again, if this happened, it was theorized that the excessive quantities of solvent and unreacted monomer accumulated in the hopper car and formed an explosive mixture.

Table 9-6   Questions/comments for Problem 9-3

Question	Comment
Are all operating directives and procedures correct and being followed?	All were being followed. However, there had been recent changes to the directives and procedures. Some of these appeared highly questionable.
Are all instruments correct?	The instruments used to monitor the stripping operation were calibrated weekly.
Are laboratory results correct?	The volatiles results were not routinely measured. It was believed that maintaining adequate temperature was all that was required to maintain volatiles control.
Were there any errors made in the original design?	Not applicable.
Were there changes in operating conditions?	Yes. See above.
Is fluid leakage occurring?	Not applicable.
Has there been mechanical wear that would explain problem?	Not applicable.
Is the reaction rate as anticipated?	Not applicable
Are there adverse reactions occurring?	Unknown changes in the catalyst might create a polymer with a less porous structure that would make volatiles removal more difficult. This could be detected by bulk density measurements. In addition, the polymer might be decomposing in the extruder forming volatile materials.
Were there errors made in the construction?	Not applicable.

3. There was a change in the adequacy of stabilization so that the polymer decomposed in the extruder. This decomposition resulted in the formation of volatile materials which then did not vent out of the extruder because of the short residence time. They were trapped in the pellet and were carried into the hopper car where they evolved from the pellet and accumulated and formed an explosive mixture.

It was clear, after considering these three hypotheses, that all of them had a common thread. The common thread was twofold. In the first place, the hydrocarbons (solvent and unreacted monomer) had to build up to the point of forming an explosive mixture. The second aspect of this hypothesis was that the explosive mixture was ignited by an ignition source. This ignition source could have been something dramatic, like a lightning strike, or something as common place as static electricity. The movement of pellets in the hopper car while it was in transit could generate sufficient static electricity to create an ignition source. This would be particularly true if the relative humidity of the air in the vapor space was very low. The problem solver used the traditional process engineering safety assumption: When dealing with an explosive mixture, an ignition source will always be found if the mixture is in the explosive range for a sufficient amount of time. Thus he only considered which hypothesis should be investigated for how an explosive mixture formed. All of the three hypotheses required that the gases accumulate in the vapor space to form an explosive mixture. The first question that the problem solver had to answer was "How much hydrocarbon had to be left in the polymer for an explosive mixture to form in the vapor space of the hopper car?"

It was necessary to develop some basic data before the problem solver could begin to assess this question. The data that he developed is shown in Table 9-7.

In order to test the hydrocarbon accumulation hypothesis, the following calculations were done to determine how much hydrocarbon would be left in the polymer entering the hopper car for an explosive mixture to accumulate in the hopper car vapor space. These calculations assumed that the polymer had been in the hopper car for such an extended period of time that equilibrium between the vapor space and the polymer was reached. The calculations also ignore the purging of the vapor space that might occur as the hopper car traveled to the customer. As such, they represent the worst case scenario. However, the case might occur if the hopper car sat stationary on the railroad track for an extended period of time with a minimal amount of wind. It was

Table 9-7 Data for evaluation

Item	Value
Hopper car capacity, lb of polymer	180000
Hopper car fill volume, %	70
Bulk density of the polymer, lb/ft^3	25
Skeletal[a] density of the polymer, lb/ft^3	56
Vapor pressure of monomer at 100°F, psia	220
Molecular weight of monomer	42
Explosive range of monomer,[b] volume %	
Lower (LEL)	2
Upper (UEL)	11

[a] Skeletal density is the density of the polymer if the particle had no voids.
[b] Since essentially the entire hydrocarbon was unreacted monomer, no consideration was given to the solvent. In addition, it should be noted that ,for gases, the volume % and mol % are identical.

assumed that the equilibrium relationship given in equation (9-7), below, was applicable for this polymer. In addition to the equilibrium relationship, material balance relationships are given in equations (9-15) to (9-18). These relationships can be used to determine the residual pounds of monomer that would be required for the vapor space to be at the lower explosive limit of 2%.

$$X_E = 295,000 \times (Y \times \pi / VP)^{1.5} \tag{9-7}$$

$$F = V + S \tag{9-15}$$

where

$\quad$ F = total monomer in the hopper car, lb
$\quad$ V = monomer in vapor space, lb
$\quad$ S = monomer remaining in polymer at 100°F, lb

$$T = Ca \times 100 / (BD \times FV)$$
$$= 180,000 \times 100 / (25 \times 70) = 10285 \text{ ft}^3 \tag{9-16}$$

$$VT = T - Ca/SD$$
$$= 10285 - 180,000/56 = 7070 \text{ ft}^3 \tag{9-17}$$

where

$\quad$ T = total volume of hopper car, ft^3
$\quad$ Ca = polymer capacity of hopper car, lb
$\quad$ BD = bulk density of polymer, lb/ft^3
$\quad$ FV = volume of hopper car filled with polymer, %
$\quad$ VT = vapor volume of hopper car, ft^3
$\quad$ SD = skeletal density of polymer, lb/ft^3

Note that VT includes the volume associated with the pores in the polymer.

$$D = MWM \times 520 / (379 \times (460 + 100))$$
$$= 42 \times 520 / (379 \times 560) = 0.1029 \text{ lb/ft}^3 \tag{9-18}$$

where

$\quad$ D = density of monomer gas, lb/ft^3
$\quad$ MWM = molecular weight of the monomer

$$V = LEL \times D \times VT / 100$$
$$= 2 \times 0.1029 \times 7070 / 100 = 14.55 \text{ lb}$$

$$X_E = 295,000 \times (Y \times \pi / VP)^{1.5}$$
$$= 295,000 \times (0.02 \times 14.7 / 220)^{1.5} = 14.4 \text{ ppm} \tag{9-7}$$

$F = V + S$

$\quad = 14.55 + 14.4 \times 180000/1,000,000 = 17.14 \text{ lb}$

$\quad = 17.14 \times 1,000,000/180,000 = 95 \text{ ppm of monomer in the incoming polymer}$

$$(9\text{-}15)$$

Thus if the polymer going to the hopper contained more than 95 ppm of monomer, the vapor space could well be equal to or above the lower explosive limit (LEL).

The problem solver investigated the current operating directives and procedures and their bases. When he did this investigation, he found that the approach to setting the new directives and procedures had been strictly empirical with no consideration to the theory and time elements of diffusion. A test run had been conducted by establishing the new conditions and rates in the polymerization and extrusion sections. When the conditions were well established, samples downstream of the extrusion and pelletizing operations were collected in an open mouth container. A conventional gas explosivity analyzer was inserted into the container immediately after the sample was collected and the percentage LEL was measured. The explosivity analyzer gives a reading of the percentage LEL. For example, if the vapor space was at the LEL, the analyzer would read 100%. Thus percentage LEL is a measure of the approach to the lower explosion limit.

In this test of X3, it was believed that measuring the vapor space immediately would give the highest value, since the monomer in the vapor phase of the open container would not have time to diffuse into the surrounding air. A single measurement was made and since the percentage LEL was less than 100%, it was concluded that the new conditions and rates would produce a product that would satisfy the DOT regulations. These regulations stated that the vapor space of products shipped in hopper cars should be below the LEL. Obviously, if the vapor space in the hopper car was below the LEL, it would not ignite.

In addition, the problem solver reviewed the bulk density measurements. The experimental polymer (X3) actually had a slightly lower bulk density than the normal polyolefin products. Thus it appeared that the product would be more porous than the conventional product. He also reviewed the current stabilizer and found that it was the same that had been in use for several years. This cursory review did not mean that he had ruled out hypotheses 2 and 3 conclusively. But it did indicate that these were unlikely to be the simplest root cause. Following the concept of taking the simplest route whether it is the root cause or a calculation technique, the problem solver developed the following theoretically correct working hypothesis.

"It is believed that the hopper car fires and product charring that occurred when shipping X3 are associated with the residual monomer that remained with the polymer when the new rates and operating conditions were used. While the vapor space above the pellets was not in the explosive range immediately after they were loaded into the hopper car, the evolution of vapors from the polymer over time allowed an explosive concentration to build up in the hopper car vapor space. While it is uncertain what the source of ignition was, it is known that static electricity is almost always present due to the movement of the polymer particles. Whether this explosive mixture ignited and to what degree a fire occurred depended on three factors:

1. "The movement of the hopper car. If the hopper car was continuously in motion, there was a high probability that some of the vapors would be replaced by air and the vapor phase concentration might be reduced to a level below the lower explosive limit.
2. "The humidity of the ambient air. If the humidity was high in the hopper car, it is unlikely that a static discharge would occur.
3. "The temperature. If the temperature of the polymer in the hopper car was elevated (due to the ambient temperature), then there is a higher probability of a major fire occurring, rather than a flash fire that chars some of the polymer on the top of the hopper car.

"In order to eliminate this problem, the production rates of X3 should be reduced to the previous levels. In addition, the new operating directives and procedures should be abandoned."

Step 4: Provide a mechanism to test the hypothesis.

Two mechanisms were provided to test the hypothesis. On the next X3 run, the rates were reduced to the previous levels. The percentage LEL was measured in the vapor space of each hopper car after 24 hr and prior to shipping. In each case, the LEL was well below 100%. There were no reports of charred polymer or fires even though the customers had been alerted to watch for such events. In addition, during the run, samples were taken of the product leaving the pelletizing section in the same fashion as discussed earlier. The percentage LEL was determined as a function of time and Figure 9-5 was used to illustrate

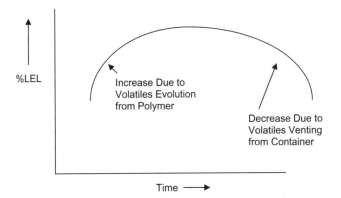

Figure 9-5   Percentage LEL vs. time.

the buildup of the percentage LEL as a function of time. Thus it was concluded that reducing the production rates was a successful test of the hypothesis.

Step 5: Recommend remedial action to eliminate the problem without creating another problem.

The recommendation to operate at reduced rates could only be considered an interim recommendation. The problem solver used some of the techniques described in Example Problem 9-1 and concluded that the only way to operate at full rates while producing this polymer was to add additional dryer residence time. He recommended that a study be initiated to determine whether the economics of producing X3 could justify either operating at reduced rates or adding a larger dryer.

Lessons Learned   There were pressures to increase the production rates of X3. It was obvious, in hindsight, that these pressures caused a blatantly flawed empirical test to be developed. The conclusion that rates could be increased was based completely on this flawed test with no theoretical calculations. This empirical test overlooked the obvious: that vapor evolution from the polymer that occurred over time might cause a maximum in the relationship between percentage LEL and time. This maximum could well be above the LEL. The risk of this approach could have been discovered if theoretical calculations had been done to determine what the maximum equilibrium concentration of monomer in the vapor phase of the sample container would be. This calculation would require that the concentration of the monomer in the polymer leaving the pelletizing section be known. The concentration of monomer in the vapor phase could then be calculated using the techniques described earlier and with the assumption that none of the monomer dispersed into the atmosphere from the container.

There are times when theoretical calculations are more accurate than laboratory results. In this case, if the concentration of monomer in the polymer could be analyzed and the equilibrium relationship was available (e.g., equation (9-7)), then the maximum equilibrium vapor phase concentration and the percentage LEL could be determined easily. This would be more accurate than the maximum percentage LEL determined by the explosivity meter. This is because the explosivity meter only measures the percentage LEL at a point in time which is not likely to be the maximum concentration.

The failure to do a detailed theoretical analysis of the proposal to increase rates by modified operating directives and procedures was costly. It gave the company adverse public publicity and adverse relationships with their customers. Their customers had never seen charred polyolefin products from this company or any other company from which they purchased material. The approach also cost problem-solving time. Instead of starting to understand the monomer removal limitation when X3 was first discovered to have a limitation, the empirical approach caused the start of problem-solving activities to be delayed. In addition, problem-solving activities were diverted to understanding the hopper car fires as opposed to eliminating the drying limitation.

The approach of asking customers to be on the watch for charred polymer might seem to be "asking for trouble." However, it was mandatory to confirm that the plant test of operating at reduced rates was truly successful. The failure to get this information would have likely resulted in a failed test, since not enough data would be present to prove or disprove the hypothesis.

NOMENCLATURE

BD	Bulk density of polymer, lb/ft^3
C	A constant referred to as the "lumped parameter constant." If it is related to heat transfer, the constant is generally taken as $U \times A$. If reaction is involved, it is simply set equal to C_R in this chapter. If drying or stripping is involved, the value K is used in this chapter.
C_R	Lumped parameter constant for reaction (reaction rate constant)
C_E	A constant that relates equilibrium of solvent in polymer to the vapor phase composition. It can be determined from experimental data or approximated by application of theoretical relationships.
C_1	A constant that is related to the type of furnace flame. It should not change unless the type of fuel changes.
C_2	Product of the heat transfer coefficient and the area of the tubes. It will only change if the heat transfer coefficient changes
Ca	Polymer capacity of the hopper car, lb
DF	Driving force or incentive for mass or heat transfer to occur
DF_R	Driving force for radiant heat transfer
dX/dt	Rate of disappearance of component X by reaction or stripping
F	Total monomer in the hopper car, lb
FV	Volume of the hopper car filled with polymer, %
K	Lumped parameter constant for drying or stripping. It is somewhat related to diffusion. Since it deals with mass transfer, it is also referenced in this chapter as a mass transfer coefficient.
LEL	Lower explosive limit, volume or mol %
$\ln T$	Log mean temperature difference or the driving force for heat transfer
P	Partial pressure of the solvent in the vapor phase

Q	Rate of heat transfer
R	Rate of change with time of the variable under study. It could be heat transfer, reaction rate, or volatile stripping
S	Monomer remaining in polymer at 100°F, lb
SD	Skeletal density of polymer, lb/ft³
T	Total volume of the hopper car, ft³
t	Amount of time in the dryer, min
T_P	Absolute temperature of the process fluid in the tubes
T_G	Absolute temperature of the gas
T_M	Absolute temperature of the tube metal
$U \times A$	Lumped parameter (C) or, for heat transfer, the heat transfer coefficient multiplied by the area
V	Monomer in vapor space of the hopper car, lb
VP	Vapor pressure of the solvent at the dryer temperature
VT	Vapor volume of the hopper car, ft³
$(X - X_e)$	Driving force for stripping
X	Actual concentration of solvent in a polymer
X	Concentration of component X
X_e	Equilibrium concentration of solvent in the polymer
X_f	Concentration of solvent in the outlet polymer
X_o	Concentration of solvent in the inlet polymer
Y	Concentration of component Y or the concentration of the solvent in the vapor phase
π	Total pressure, psia

10

APPLICATION TO UNSTEADY STATE

10.1 INTRODUCTION

While most process engineering courses deal with design and operations at
steady state conditions, and most process operators target to achieve steady
state, industrial processes operate at constantly changing conditions. The differences between steady and dynamic conditions can frequently be ignored.
However, an approach to considering unsteady operations will be of value to
every problem solver. There are multiple instances where unsteady state operations must be considered. Startups are never steady state; this is probably
the most challenging phase of problem solving. This is often an area where
the intuition of an experienced operator or engineer is more valuable than
a detailed problem-solving activity. However, there are examples where a
frequently encountered startup activity that involves a problem (reactor
startup, for example) can be approached using some of the principles outlined
in this book.

Batch operations are always dynamic. If problems are uncovered during a
batch operation, typical process engineering calculations based on steady state
operations will rarely be of any value. Thus the industrial problem solver needs
to have tools to allow him to approach these dynamic operations. Problem
10.1 is a typical example of a problem within a batch operation.

Unsteady state upsets can often provide valuable information. For example,
short term upsets caused by feed impurities can be extrapolated to steady state
values using techniques that relate the time of the upset to the actual reactor

Problem Solving for Process Operators and Specialists, First Edition. Joseph M. Bonem.
© 2011 John Wiley & Sons, Inc. Published 2011 by John Wiley & Sons Inc.

residence time. This will allow the problem solver to determine the steady state impact of these impurities. This is the impact on the process that would occur if the short term impurity concentration in the feed were at the upset level consistently. This information can be used to determine the specification of the impurity in the feed.

10.2   APPROACH TO UNSTEADY STATE PROBLEM SOLVING

The most difficult aspect of solving unsteady state problems in an industrial environment is the balance between formulating technically correct hypotheses and the need for expediency. This was referred to in Chapter 3 as optimum technical depth. Because of the difficulty of determining this balance, one of two extremes is often present. In one of these extremes, the problem solver begins developing a highly sophisticated dynamic model that requires a powerful computer to solve the differential equations. As the work slowly progresses, management becomes impatient and cancels the project. At the other extreme, the problem solver often gives up and uses intuitive methods, rather than developing a simplified dynamic model. Optimum technical depth is the approach used to achieve a balance between these two extremes. This balance will be a strong function of the environment surrounding the problem. For example, during a startup, a balance slanted toward intuition is likely the best approach. On the other hand, a chronic problem which does not have any apparent answer will require a more sophisticated approach. The following paragraphs present guidelines for approaching unsteady state problems.

If the problem requires an immediate recommendation, such as during a startup, the best approach will be to rely on the instinct of an operator or experienced startup engineer. Medical research has shown that our minds work by both instinct and logical reasoning; obviously, instinct is faster than development of a detailed logical path. Minds conditioned by years of experience with unsteady state operations such as a startup will react instinctively to provide an immediate response, as opposed to having to think about the situation.

The basic concept used when considering unsteady state operations is equation (10-1) shown below:

$$AD = I - O - RD \tag{10-1}$$

where

AD = rate of accumulation (this could be accumulation of anything: level, heat, or reactant)

I = inflow of material or heat

O = outflow of material or heat

RD = removal/addition of heat, or formation/destruction of material by reaction

This relationship can be used to as a building block for all unsteady state considerations from something as simple as the change of level in an accumulator to the development of complicated dynamic models. The main goal of this chapter is the development of simple but accurate dynamic models.

If it is considered desirable to build a dynamic model of the unsteady state process, all physical components of the system must be considered. For example, with steady state, the heat capacity of the process vessel is correctly never considered. This is because the vessel wall and the vessel content temperatures remain essentially constant. However, for unsteady state heat balances, the heat capacity of the vessel (both the walls and contents) must be considered. The heat capacity of the vessel walls will often be a moderating influence. This moderating influence will cause the rate of temperature change to be less than that calculated if the complete heat capacity of the vessel is ignored. The dynamic model should be kept simple by use of idealized mixing patterns, use of lumped parameter constants (such as overall heat transfer coefficients), and an assumption of uniform metal temperature. The two idealized mixing patterns are either a perfectly mixed vessel or "plug flow." Plug flow involves an assumption that every particle has exactly the same residence time in the vessel. An example of a plug flow mixing pattern is a fixed bed dryer or reactor where the velocity is so low that a streamline flow pattern is developed. In this case, the residence time for every element is the vessel fluid holdup divided by the volumetric flow rate. The opposite mixing pattern is the perfectly mixed vessel. In a perfectly mixed vessel, the contents of the vessel are at the concentration of the outlet fluid. An example of a perfectly mixed vessel is a reactor used to produce polypropylene from propylene. The pure propylene enters the reactor, where about 50% is converted to polymer. The outlet concentration is 50% by weight polypropylene. Since the reactor is a perfectly mixed vessel, the polypropylene concentration at any point in the vessel is also 50% by weight.

If reaction is involved, it should be recognized that there are two models that can be used to simulate an unsteady state reaction. Traditional process engineering indicates that the reaction rate is proportional to the concentration of the reactant. That is, the rate of reaction will increase as the concentration of the reactant in the vessel increases. This has been referred to as the "slow response model." For some reactants, such as impurities in a polymerization reactor, the impurity reacts as soon as it enters the reactor. This "fast response model" is easier to develop because it allows an engineer to assume that the concentration of the impurity in the reactor approximates zero. In order to determine whether the traditional model or the "fast response model" should be used, the problem solver must understand the technology. As a general rule, the "fast response model" will apply only to reaction impurities in a polymerization reaction. Most other reaction modeling will be more accurate if the traditional process engineering approach is utilized. Regardless which model is used, short term upsets and fundamental models of stirred tank or plug flow reactors can be used to estimate the full impact of an impurity.

10.3 EXAMPLE PROBLEMS

The problem-solving techniques associated with unsteady state operations are illustrated by the actual case histories described in the following paragraphs. Example Problem 10-1 illustrates the use of these techniques for solving a problem associated with a batch reactor which experienced a temperature "runaway" (such a rapid temperature increase that the emergency devices were activated). Problem 10-2 illustrates the use of these techniques to extrapolate short term reaction upsets caused by an impurity to steady state conditions.

EXAMPLE PROBLEM 10-1

A batch reaction was carried out in a reaction vessel with a cooling water jacket. The reaction was initiated when the primary reactant was added to the reaction vessel. The primary reactant was dissolved in hot hexane in a small vessel and then pressured into the reaction vessel with nitrogen as fast as possible. While the technology had been developed in the research facility of a major petrochemical company, the actual manufacturing of the material was being done by a contract manufacturer. His operation was designed and operated on an exceptionally low budget. There were no flow instruments to measure and control the rate of addition of the reactant. The average rate of reactant addition was obtained knowing the amount of material in the small vessel and the time that it took to add it to the reaction vessel.

The exothermic (heat generated) reaction had been conducted successfully many times. However, during a recent batch a rapid increase in temperature was experienced as the primary reactant was added to the reactor. The rapid temperature increase was so fast and of such a magnitude that the safety release system on the reactor was activated. The actual manufacturing location was located several hundred miles away from the technology center. Operations personnel decided to solve the temperature runaway problem without asking for help from the technology center. In order to eliminate the rapid increase in temperature, operations personnel decided to add the reactant at a very slow rate. The next batch was produced without a temperature runaway. However, the product produced in this fashion did not meet the product morphology specifications. Morphology includes such physical attributes as particle diameter, porosity, and surface area.

After this batch failed to meet the morphology specifications, the operations personnel requested help from the technology center. The problem solver was charged with the responsibility of determining how to conduct this reaction with an 80 to 90% probability that a second temperature runaway would not occur and so that the product morphology would be satisfactory.

The problem solver began accumulating technology information. It was known that the primary reactant reacted instantly when it entered the reactor (fast response model). It was desirable to add the reactant as fast as possible for two reasons. As indicated earlier, the morphology of the product was adversely impacted by a slow addition rate of the reactant. In addition, the heat of reaction provided heat input to raise the reactor temperature rapidly from 149°F to 185°F.

The heat of reaction was unknown. It was not determined in the laboratory during the development part of the project. Actually, the laboratory chemists believed that there was no heat of reaction. The small scale reactor had an inherently high heat transfer area-to-volume ratio. Laboratory reactors will almost always have a higher heat transfer area-to-volume ratio than commercial size reactors. This always creates a potential problem in scaleup to a commercial-size process. In addition, this laboratory scale reactor was surrounded by a constant temperature sand bath. This combination essentially eliminated the chance to observe any heat of reaction in the laboratory. The nature of the reaction was such that the heat of reaction could not be determined by classic literature-based approaches. In addition, time and financial constraints did not allow determination of this variable from laboratory data.

The problem solver began using the five-step problem-solving approach described earlier, as is shown below:

Step 1: Verify that the problem actually occurred.

Verification of this problem was easy since the hexane vapor from the safety release system on the reactor (a rupture disc) condensed in the air and "rained" on a local Veteran's Day parade. There were no injuries, but several uniforms and band instruments required cleaning and reconditioning. In addition, a review of the reactor temperature and pressure data indicated that the rupture disc did not release prematurely. That is, the pressure and temperature had really gotten out of control.

Step 2: Write out an accurate statement of what problem you are trying to solve.

The problem statement developed by the problem solver was as follows:

An exothermic reaction conducted in a batch reactor had a temperature runaway even though the same reaction had been conducted many times previously with no temperature runaway. In a subsequent batch it was demonstrated that the temperature could be controlled by a slow rate of addition of the primary reactant. However, it is necessary to add the primary reactant at a rapid rate in order to obtain the desired product morphology. Determine what caused the batch reactor to have a temperature runaway In addition, develop procedures that ensure that there is at least an 80 to 90% confidence level that this temperature runaway will not reoccur when producing a product that meets the morphology specifications.

Step 3: Develop a theoretically sound working hypothesis that explains as many specifications of the problem as possible.

Table 10-1 shows the questions from Chapter 6 along with appropriate comments.

A review of the comments associated with these questions and a review of all previous runs indicated several points of interest that would require additional considerations. These were as follows:

- While the batch size was the same, the rate of reactant addition varied greatly from run to run. As indicated earlier, this rate of addition was only determined after completion of the addition since there were not flow meters on the transfer line. There were no procedures given for the rate of reactant addition. Some operations personnel pressured the drum used to mix the reactant and hexane with nitrogen prior to opening the

Table 10-1 Questions/comments for Problem 10-1

Question	Comment
Are all operating directives and procedures being followed?	For the batch that had a temperature runaway, all appeared to have been followed. However, there were no guidelines given for how fast to add the primary reactant.
Are all instruments correct?	The instruments had allegedly been calibrated.
Are laboratory results correct?	The product morphology on the second batch was confirmed by two independent techniques.
Were there any errors made in the original design?	Yes. The laboratory chemists indicated that there was essentially no heat of reaction based on their studies. In hindsight, this was obviously an error.
Were there changes in operating conditions?	Yes. Rate of addition of reactant was highly variable even though batch sizes were the same.
Is fluid leakage occurring?	Not applicable.
Has there been mechanical wear that would explain problem?	This is possible, but it would cause the heat removal capability to decrease. This was not occurring based on measured coolant flows and temperatures.
Is the reaction rate as anticipated?	Reaction rate highly dependent on reactant rate of addition.
Are there adverse reactions occurring?	Adverse reactions were likely the cause of poor morphology. However, this is exceptionally complicated chemistry. Thus solving the morphology problem while maintaining slow addition of reagents is unlikely.
Were there errors made in the construction?	Not applicable. Some batches worked fine.

transfer valve. Others simply started the nitrogen flow to pressurize the drum and then opened the transfer valve. There were times when the mixing drum was hotter than it was at other times, which impacted the solution viscosity and would affect the rate of reactant transfer.

- While the laboratory chemists reported that the heat of reaction was insignificant, there had to be a significant heat of reaction to cause the vessel contents to increase from 149°F to 185°F in a short period of time. If the heat of reaction was known or could be determined, it would be possible to estimate the rate of temperature increase for any batch size and/or rate of reactant flow.

- In addition to the problem batch, there were several other runs where the temperature increased almost as rapidly, although the reactor temperature did not become unstable. The only difference in these runs was the rate of reactant addition.

While it could be theorized at this point that the temperature runaway was due to the rate of reactant addition, this theory would only allow one to specify that an addition rate equivalent to that of the first failed batch would cause a temperature runaway. It would not allow determination of the maximum rate of reactant addition. A more fundamental approach was required to allow specification of the target reactant addition rate.

The problem solver decided to approach the problem using a fundamental approach. He felt that if he could express the factors involved in the temperature runaway mathematically, he could understand better how to prevent it. A temperature runaway occurs when the rate of heat generated is greater than the maximum rate that heat can be removed. This can be expressed mathematically as follows in equations (10-2) through (10-4):

$$Q_g > (Qr)_{max} \qquad (10\text{-}2)$$

where

Q_g = rate of heat generation, BTU/hr
$(Q_r)_{max}$ = maximum rate of heat removal, BTU/hr

$$Q_g = \Delta H_R \times R \qquad (10\text{-}3)$$
$$(Q_r)_{max} = U \times A \times (T_R - T_C) \qquad (10\text{-}4)$$

where

ΔH_R = heat of reaction, BTU/lb
R = rate of reaction (for the case of an instantaneousreaction, it is the rate of reactant addition, lb/hr)
U = heat transfer coefficient, BTU/hr-°F-ft^2
A = heat transfer area, ft^2

T_R = temperature of the reactor, °F

T_C = minimum temperature of the coolant, °F

Note that equation (10-4) has been simplified from the traditional heat transfer equation that involves use of a logarithmic relationship. This is valid for this particular case since the coolant is flowing through the reactor with a minimal increase in temperature. The temperature to be used is the minimum coolant temperature that can be obtained. It is assumed that the control system will react to provide the minimum coolant temperature possible. In addition, because the reactor vessel is well mixed, the reactor temperature throughout the vessel is constant at any point in time.

For a batch reactor, equations (10-3) and (10-4) can be combined and a heat accumulation term can be added to take into account the unsteady state nature. The amount of heat accumulated in the vessel will equal the heat of reaction minus the heat removed. Any heat accumulation will cause an increase in the temperature of the reactor. The increase in temperature will depend on the heat capacity of the vessel contents and on the vessel itself. Heat capacity is basically the weight of material multiplied by its specific heat. The equations below were developed to represent this situation.

$$AC = Q_g - (Q_r)_{max} \tag{10-5}$$

$$AC = \Delta H_R \times R - U \times A \times (T_R - T_C) \tag{10-6}$$

$$AC = W \times C_P \times dT/dt \tag{10-7}$$

where

AC = heat accumulation, BTU/hr

W = weight of material (metal, water, reactants), lb

C_P = average specific heat of material, BTU/lb-°F

dT/dt = rate of temperature rise, °F/hr

Since the reaction of the reactant being pressured into the reactor is instantaneous, the heat generated ($\Delta H_R \times R$) depends only on the rate of reactant addition. Thus equations (10-6) and (10-7) can be combined and modified to simplify solving for the heat of reaction.

$$\Delta H_R = (W \times C_P \times dT/dt + U \times A \times (T_R - T_C))/R \tag{10-8}$$

In the above equation, U and A can be estimated from physical dimensions and typical vessel heat transfer coefficients. Since all previous batches and future plans called for use of the same vessel, obtaining exact values of U and A was not important. The product of UA was considered a "lumped parameter constant."

Equation (10-8) was used to determine the heat of reaction for all previous runs. Since a reasonably constant heat of reaction was calculated over a wide range of reactant addition rates (R), this approach seemed valid. After determining the heat of reaction using data from the batch runs, calculations were done to determine the maximum rate of reactant addition. Referring back to equation (10-2), a typical approach is to provide a 10% safety factor, so that the maximum rate of heat removal is 10% greater than the maximum rate of heat generated. Thus equation (10-2) can be modified as follows:

$$1.1 \times Q_{g} < Q_{r\,max} \tag{10-9}$$

In order to estimate the maximum rate of reactant addition, equation (10-9), along with equation (10-10), shown below (which is rearrangement of equation (10-8)), were used.

$$dT/dt = (R \times \Delta H_{R} - U \times A \times (T_{R} - T_{C}))/(W \times C_{P}) \tag{10-10}$$

Equation (10-10) was used to estimate the maximum rate of reactant addition (R) that would allow the temperature to rise from 149°F to 185°F without causing a temperature runaway. This maximum rate of addition was determined using an iterative procedure along with a spread sheet. The calculated relationship between the reaction temperature and time for the maximum rate of addition is shown in Figure 10-1. After the maximum rate was determined, it was reduced by 10% to be consistent with the requirements shown in equation (10-9).

Using this approach, it was estimated that the maximum rate of reactant addition was 25% less than that which was utilized in the batch on which a

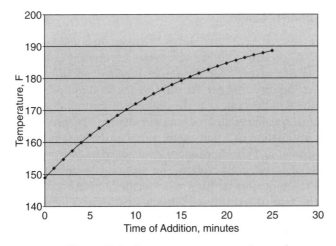

Figure 10-1 Reactor temperature vs. time.

temperature runaway occurred. The problem solver developed the following working hypothesis:

> It is theorized that the temperature runaway was caused by the rapid addition of the primary reactant. This addition rate was so high that the temperature could not be controlled, even when the coolant was at the maximum rate. Calculations indicate that if the primary reactant rate addition was reduced by 25%, the reaction could be controlled.

Step 4: Provide a mechanism to test the hypothesis.

Developing a mechanism to test the hypothesis was easy from an engineering standpoint. As described above, the maximum rate of reactant addition to achieve a peak temperature of 185°F was easily estimated. However, convincing the operations personnel to add the reactant at this rate was difficult since they had recently had good success controlling the temperature while adding the reactant at a much lower rate and allowing the reactor to slowly heat up. As indicated earlier, this resulted in a product with poor morphology. In order to convince them to test the hypothesis, it was necessary to give them a complete explanation of the importance of the morphology of the product and to stand by the reactor when the reactant was added at a rapid rate. The problem solver had estimated the nitrogen pressure that would be required to transfer the reactant to the drum at the desired rate. He had also calculated the rate of level change in the reactant storage drum that would result if the reactant were added at the correct rate. The test was successful in producing a product with good morphology and the temperature rose, as predicted, without a temperature runaway.

Step 5: Recommend remedial action to eliminate the problem without creating another problem.

The hypothesis test (step 4) proved that the reactant could be added at a high rate. In order to prevent another temperature runaway, additional controls were considered to assure that the target reactant addition rate was maintained, Since, as indicated earlier, the only means used by operations personnel for knowing the addition rate was the amount of time that it took to add the reactant, a more specific technique was mandatory. The project did not have funds for the installation of a flow meter, nor was the required shutdown time immediately available. Three approaches were used to ensure that the reactant was being added at the correct rate. The pressure to be reached and maintained on the reactant storage drum was specified in the operating procedures. In addition, the rate of level change that would be anticipated was specified. As a final check, the curve shown in Figure 10-1 was provided and was to be used to monitor the rate of temperature rise in the first few minutes. For example, if the initial temperature increase was too fast, it would indicate that, in spite of the other techniques, the rate of addition was too fast. The rate of addition would then be reduced.

Lessons Learned   Although it has nothing directly to do with problem solving, one of the lessons learned was that it is very difficult to determine heat of reaction from laboratory experiments that are not specifically designed to obtain this variable. A typical laboratory experiment will be designed to obtain reaction rate variables or product quality attributes. The best way to do this will be at a constant temperature. This is usually done by utilizing small equipment which has a high area to volume ratio and is surrounded by a constant temperature bath. Within this facility, it will be impossible to determine the heat of reaction or even to notice whether there is a heat of reaction or not.

The utilization of the unsteady state relationship, equation (10-1), for heat accumulation proved to be a reliable tool for determining the heat of reaction. As indicated, the key to being able to successfully use this equation is the full inclusion of all of the heat content (vessel and reactants) of the process under consideration. The qualification that time and money were not available to determine the heat of reaction in the laboratory is often the case in industrial problem solving. In order to obtain the heat of reaction in the laboratory, it would have taken time and funds to build a laboratory reactor specifically designed to obtain heat of reaction data.

In problem solving, the engineering calculations are often the easiest part of the job. Convincing either operations management or the hourly work force of operators and mechanics that the recommended solution is correct is often more difficult. The process of convincing the operations organization to adopt the answer will be easier if they are convinced that the problem solver is there with them to assist them if something was wrong with the calculations. Thus in this example problem, the problem solver was present when the operator added the reactant at the calculated rate. While things went as planned, the operator felt much more comfortable in setting the pressure to give the high rate of flow since the problem solver was present with him.

EXAMPLE PROBLEM 10-2

A continuous stirred tank reactor (CSTR) in a polymerization plant was plagued by short term "loss of reaction" events. While these events were small in magnitude, they created product quality upsets as well as a loss of catalyst efficiency. This type of reactor is highly agitated and can be treated for all practical purposes as a perfectly mixed vessel.

These loss of reaction upsets appeared to be associated with spikes of carbon dioxide (CO_2) in the monomer feed. The reactor residence time was approximately 3 hr. The short-term spikes of CO_2 lasted for only 30 to 60 min.

Some batch data was available from the laboratory indicating that the CO_2 specification of the monomer feed currently at 10 ppm needed to be reduced. However, management would like to be convinced that the laboratory data was consistent with the plant experience. Management also believed that there was probably another impurity that was not being analyzed that was causing the problem. This belief was based on previous experience in another plant that utilized a similar, but not identical, catalyst system. Management was also concerned that a project to reduce the CO_2 specification and eliminate spikes would take several months to implement and require significant investment. Thus they did not want to proceed with the project unless there was a high probability of success.

The problem solver was asked to determine if these upsets were really caused by feed impurities, what the likely feed impurity was, and the impurity's real impact on the process. It was believed that this approach would lead to determining and setting realistic impurity specifications on the feed monomer.

The problem solver approached the problem using the 5-step problem-solving technique as shown in the following paragraphs.

Step 1: Verify that the problem actually occurred.

Verification of this problem was done through the use of two different techniques. The polymer was analyzed to determine the concentration of catalyst in the polymer leaving the reactor. From this result, the catalyst efficiency was calculated and it was clear that, during CO_2 spikes, the catalyst efficiency did decrease. However, this approach only gave an average catalyst efficiency at any point in time. Since the reactor had a 3-hr residence time and the impurity spikes lasted only 30–60 min, this average represented 2–2.5 hr of normal operation and 0.5–1 hr of upset operations. In addition to this analytical approach, the calculated production rate (as determined by the process control computer using a heat balance) was used to determine the instantaneous production rate. Both of these techniques confirmed the loss of reaction that occurred about the same time that the CO_2 content of the feed increased.

Step 2: Write out an accurate statement of what problem you are trying to solve.

The statement written by the problem solver was as follows:

> Frequent loss-of-reaction episodes are occurring in a polymer plant. These episodes cause a decrease in average catalyst efficiency as well as inconsistent product quality. These episodes appear to occur at the same time as short term spikes of CO_2 concentration in the monomer feed. Determine the cause of the frequent "loss of reaction" episodes. Any conclusions based on plant data must be consistent with available laboratory data on the impact of CO_2 on the polymerization reaction. This evaluation should also indicate whether or not there is

any indication that another impurity besides CO_2 is present. Based on this data, the monomer feed specification should be reviewed.

Step 3: Develop a theoretically sound working hypothesis that explains as many specifications of the problem as possible.

In order to fully assess the possible causes of the upsets, the questions shown in Chapter 6 were utilized and Table 10-2 was developed.

The above comments tended to confirm the suspicion that the upsets were caused by the presence of CO_2. However, there was still a residual suspicion that there was another impurity that was causing the problem. The problem solver determined that if the batch laboratory data could be shown to be consistent with plant data, this suspicion would be removed. The batch laboratory

Table 10-2   Questions/comments for Problem 10-2

Question	Comment
Are all operating directives and procedures being followed?	All appeared to be correct and being followed.
Are all instruments correct?	The instruments had allegedly been calibrated. In addition, the heat balance was checked by manual calculations.
Are laboratory results correct?	Yes. The x-ray machine used to determine the catalyst concentration was cross checked with another machine. No other impurities besides CO_2 were observed on continuous analyzers during upsets. In addition, samples of the feed taken during upsets and analyzed in the laboratory confirmed the presence of CO_2.
Were there any errors made in the original design?	No.
Were there changes in operating conditions?	No.
Is fluid leakage occurring?	The potential for reaction quench fluids leaking into the reactor was checked and it was concluded that they were not leaking into the reactor.
Has there been mechanical wear that would explain problem?	No.
Is the reaction rate as anticipated?	Yes, except during upsets.
Are there adverse reactions occurring?	Yes. During upsets the product quality attributes decreased.
Were there errors made in the construction?	Not applicable.

data was developed by blending CO_2 into the monomer to the desired experimental level and then conducting a polymerization. In order to develop a comparison between laboratory data and commercial data, a technique was required to allow extrapolation of short term plant upset data to steady state operations. The steady state operation values would allow comparison to the batch laboratory data. After some literature research, the problem solver developed the four important concepts described in the next few paragraphs.

In most polymerization reactions, highly reactive impurities such as CO_2 react immediately upon entering the reactor. That is, the rate of catalyst deactivation is not dependent on the concentration of the impurity in the reactor, but depends on the concentration of the impurity in the feed. As discussed earlier, this is known as the "fast response model." This is the first important concept.

The second concept is that for industrial problem solving, the polymerization reaction can be represented by simple first order kinetics with respect to the catalyst and monomer concentration. The equation shown below was utilized to approximate the polymerization kinetics in a reactor with a fixed volume.

$$R = k \times C \times M \qquad (10\text{-}11)$$

where

R = polymerization rate or monomer consumption rate, lb/hr
k = kinetic rate constant including reactor volume, ft^6/hr-lb
C = catalyst concentration in the reactor, lb/ft^3
M = monomer concentration in the reactor, lb/ft^3

Note that the units of the kinetic constant (ft^6/hr-lb) include a fixed reactor volume. The absolute value of the kinetic constant is not important since, as will be shown later, all that is considered is the relative change in this constant.

The third concept involves the modification of equation (10-1) to include a catalyst deactivation term and development of an expression for catalyst concentration in the reactor. The rate of accumulation of catalyst in the reactor can be expressed as shown in equation (10-12), below:

$$A_C = I_C - O_C - D_C \qquad (10\text{-}12)$$

where

A_C = rate of accumulation of catalyst in the reactor, lb/hr
I_C = inflow of catalyst, lb/hr
O_C = outflow of catalyst, lb/hr
D_C = deactivation of catalyst, lb/hr

Based on the "fast response model," any CO_2 entering the reactor will immediately deactivate some of the catalyst. This relationship can be expressed as follows:

$$D_C = Z \times B \qquad (10\text{-}13)$$

where

Z = ratio of catalyst deactivated to CO_2, lb/lb
B = rate of CO_2 entering reactor, lb/hr

Equation (10-12) then becomes:

$$A_C = I_C - O_C - Z \times B \qquad (10\text{-}14)$$

If the flow of CO_2 continues for an extended period, a steady state will be reached. At steady state, by definition, there is no accumulation of catalyst. Therefore

$$A_C = 0 \qquad (10\text{-}15)$$
$$O_C = I_C - Z \times B \qquad (10\text{-}16)$$

This steady state value will be the full impact of the particular rate of CO_2 entering the reactor. While the deactivation ratio (Z) is of theoretical interest, it can be eliminated from pragmatic considerations, as described later.

As indicated in equation (10-11), the important variable is the catalyst concentration in the reactor. Equation (10-14) can be transformed into equation (10-17) that uses the catalyst concentration by the appropriate substitutions.

$$dC/dt = (I_C - O_C - Z \times B)/V \qquad (10\text{-}17)$$

where

dC/dt = rate of catalyst concentration change in lb/ft³-hr
V = volume of the reactor, ft³

Since the rate of catalyst concentration change in the reactor can be determined from equation (10-17), equation (10-18), shown below, can be used to determine the catalyst concentration at any point in time.

$$C_i = C_0 - (dC/dt) \times t \qquad (10\text{-}18)$$

where

C_i = catalyst concentration in the reactor at any point in time, lb/ft^3

C_0 = initial catalyst concentration in the reactor, lb/ft^3

t = elapsed time from the start of the upset to the time of interest

The fourth concept involves developing a relationship between the concentration of active catalyst in the reactor at any point in time, the initial concentration, and the final concentration. This relationship assumes that a step change is made to an equilibrium condition and that the step change is continued until steady state is reached.

This relationship will allow translation of upset data into steady state for comparison to the batch laboratory data. For a perfectly mixed CSTR, the concentration of a component at any point in time after a step change is made can be expressed as shown in equation (10-19). In the specific case under consideration, the component is the catalyst and the step change is an increase in the CO_2 concentration in the feed.

$$(C_i - C_o)/(C_f - C_o) = 1 - e^{-T} \tag{10-19}$$

where

C_i = concentration of catalyst in the reactor at any point in time

C_o = concentration of catalyst in the reactor at time = 0

C_f = concentration of catalyst in the reactor after an infinite amount of time

T = number of reactor displacements; this is simply the time after a step change divided by the average residence time

An examination of equation (10-19) will show that after three reactor displacements, or 9 hr if the residence is 3 hr, that 95% of the final change has been achieved. That is, $(C_i - C_o)/(C_f - C_o) = 0.95$.

As indicated in equation (10-13), the amount of catalyst deactivated by each pound of CO_2 is not known and was set equal to Z. However, the instantaneous reaction rate is directionally proportional to the catalyst concentration in the reactor (assuming a constant monomer concentration) regardless of the value of Z. Thus the reaction rate can be substituted for the C values in equation (10-19). Equation (10-19) then becomes:

$$(R_i - R_o)/(R_f - R_o) = 1 - e^{-T} \tag{10-20}$$

where

R_i = reactor production at any point in time

R_o = reactor production at time = 0

R_f = reactor production after an infinite amount of time

T = number of reactor displacements at any point in time

Thus if reactor production at time $= 0$ (R_o), the reactor production at any point in time (R_i), and the number of reactor displacements at any point in time (T) are known, the steady state impact (R_f) of a step change in impurity level can be estimated. The number of reactor displacements is the time, since the step change in impurity level occurred divided by the reactor residence time. The use of equation (10-20) also eliminates some of the concerns associated with theoretically imperfect derivation of equations (10-17) and (10-18).

Unfortunately, the impurity upsets observed by the problem solver did not occur as step changes. He simulated the approximate sinusoidal curve of concentration versus time that often occurred during upsets as a step change. A typical impurity curve, along with the technique used to simulate a step change, is shown in Figure 10-2. The problem solver recognized that selecting the magnitude of the upset level to use was somewhat subjective. He tried to select it so that the areas below and above the increasing part of the actual curve were equal. He also used the peak in the concentration versus time curve as the termination point of the upset (R_i or C_i).

The problem solver used this four-concept technique because it provides a consistent method for evaluating upsets of varying magnitudes and durations. For example, an upset where the CO_2 concentration rose from 0 to 25 ppm and lasted for 15 min might have less actual impact than an upset where the CO_2 concentration rose from 0 to 5 ppm and lasted for 45 min. However, the projected steady state impact of the first upset was greater. It was obvious that a steady state concentration of 25 ppm of CO_2 in the monomer feed would have a much larger impact that 5 ppm. However, the exact magnitude of this difference could not be known without using the techniques described earlier.

In order to completely formulate the working hypothesis, several reaction upsets apparently caused by CO_2 impurities of varying severity and duration were followed and the techniques discussed above were used to extrapolate to the steady state effect. These data, along with the batch laboratory data, are

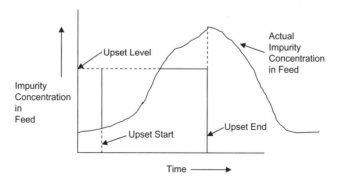

Figure 10-2   Simulation of actual change.

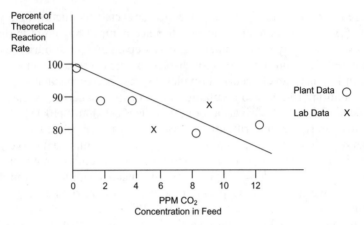

Figure 10-3   Projected steady state reaction rate vs. CO_2 concentration.

shown in Figure 10-3. The problem solver then developed the following working hypothesis based on the data shown above:

It is theorized that the short term reaction upsets are due to the presence of CO_2 in the reactor feed. The plant data is consistent with laboratory data that indicates that CO_2 is a potent reactor impurity. The reaction upsets are likely to cause product quality upsets due to the unsteady nature of the reactor during these transients. In addition, the presence of a potent reactor impurity is anticipated to affect polymer quality attributes. There were no indications of any other impurities in any of the upsets that were followed.

Step 4: Provide a mechanism to test the hypothesis.

In order to test this hypothesis, a technique had to be developed to ensure that the reactor feed was free of CO_2 for an extended period of time. If no reaction upsets occurred during this time period, agreement was reached that this would be considered a successful test and proof of the hypothesis. While it would involve additional operating cost, it was agreed to operate the monomer production facilities in a fashion that would limit the probability of CO_2 upsets. While this additional operating cost was not acceptable in the long range, it was acceptable for a one-month trial. During this one month period there was no measurable CO_2 present in the monomer feed, nor were there any reaction upsets. Thus this appeared to be a successful test and the working hypothesis was proven.

Step 5: Recommend remedial action to eliminate the problem without creating another problem.

Based on the successful test that proved the hypothesis, the following remedial actions were taken:

- An evaluation was made to determine if the specification should be reduced from the existing 10 ppm to a lower level. Because of the high cost of the catalyst and the fact that operating at 10 ppm would cause the reaction rate to decrease from 100% of theoretical to about 80% of theoretical (see Fig. 10-3), it was concluded that the specification should be reduced to 1 ppm.

- Additional facilities were installed to allow reduction of the specification from 10 ppm to 1 ppm and to eliminate the occurrences of impurity upsets.

- A new, highly accurate continuous analyzer to monitor the reactor feed for CO_2 was installed.

- Contingency plans were developed for the possibility of a CO_2 upset. These plans included shutting down the reactor if the concentration reached such a high level that serious product quality problems would be encountered.

Lessons Learned   While this problem might be considered an isolated example that has minimal application to the industrial world, it should be recognized that upsets do occur on a frequent basis. These upsets are often ignored because they are only short term. When they are considered, it is often with a minimal amount of data analysis and wrong conclusions are often reached.

In the real example problem, pursuing the presence of another impurity besides CO_2 would have been an easy trap to fall into because of the presence of previous experience and the persuasiveness of the proponent of the idea. However, it would have been the wrong route to pursue. It was only when data was developed and analyzed that it became obvious that the impurity upsets were associated with CO_2. The problem also illustrates the value of quantitative data. The relationship between the concentration and reaction kinetics (Fig. 10-3) allowed determination of the optimum monomer impurity specification.

The problem illustrates the types of short cuts that can be taken in an industrial environment. While, theoretically, polymerization reaction kinetics involve several steps such as catalyst activation, polymerization initiation, polymerization propagation, and polymerization termination, there is no need to include all of these reactions. Essentially, all polymerization reactions can be modeled by simple first order kinetics with respect to the catalyst and monomer concentration. In addition, since in this specific problem the reactor volume is fixed, there is minimal need to include it in the kinetic relationship. What is important is the model to be used for the reaction of the impurity (fast response or slow response) and the reactor fluid flow (CSTR or plug flow). In this case, the problem description includes the idea that the impurity CO_2 reacts immediately as it enters the reactor and thus can be simulated with a fast response model. The problem description also indicated that the reactor was a CSTR.

The fact that the impurity concentration versus time relationship was simulated as a step change was a decision made by the problem solver. It would have been possible to simulate this as a sinusoidal relationship. However, it is doubtful that this would have improved the accuracy of the solution and it would have increased the complexity of the approach significantly.

10.4   FINAL WORDS

The primary purpose of this chapter is to illustrate the need to both have and apply techniques for the unsteady state. While it may be unlikely that the problem solver will have problems identical to the examples provided, it is a certainty that he will encounter problems that involve unsteady state. The engineering concept given in equation (10-1) is simple.

$$AD = I - O - RD \tag{10-1}$$

The techniques required to apply the equation are more complex. However, the problem solver must not be overcome by the complexity. Like all problems and/or projects, the complexity must be approached one step at a time. For example, questions must be asked such as what is the input; what is the output; and how can the reaction, addition, or removal of heat be estimated. The benefit of doing the work required to utilize the techniques described is that quantitative problem solutions are possible. For example, in Example Problem 10-1, it was possible to specify the rate of reactant addition based on the data analysis.

The two example problems are actual problems. The time pressures to develop solutions rapidly were present. But there was also a need to work the problem right the first time as opposed to using multiple "trial and error" attempts often characterized by the phrase "We have got to try something."

NOMENCLATURE

A	Heat transfer area, ft^2
A_C	Rate of accumulation of catalyst in the reactor, lb/hr
AC	Heat accumulation, BTU/hr
AD	Rate of accumulation. This could be accumulation of anything: level, heat, or reactant
B	Rate of CO_2 entering reactor, lb/hr
C	Catalyst concentration in the reactor, lb/ft^3
C_f	Concentration of catalyst in the reactor after an infinite amount of time, lb/ft^3
C_i	Catalyst concentration in the reactor at any point in time, lb/ft^3
C_0	Original catalyst concentration in the reactor, lb/ft^3

C_p	Average specific heat of material, BTU/lb-°F
D_C	Deactivation of catalyst, lb/hr
dC/dt	Rate of catalyst concentration change in lb/ft³-hr
dT/dt	Rate of temperature rise, °F/hr
I	Inflow of material or heat
I_C	Inflow of catalyst, lb/h
k	Polymerization kinetic rate constant including reactor volume, ft⁶/hr-lb
M	Monomer concentration in the reactor, lb/ft³
O	Outflow of material or heat
O_C	Outflow of catalyst, lb/hr
R	Rate of reaction. For the case of an instantaneous reaction, it is the rate of reactant addition, lb/hr. For a polymerization reactor it is the polymerization rate or monomer consumption rate, lb/hr.
R_f	Reactor production after an infinite amount of time
R_i	Reactor production at any point in time
R_o	Reactor production at time = 0
RD	Removal/addition of heat, or formation/destruction of material by reaction
T	The number of reactor displacements. This is simply the time after a step change divided by the average residence time.
T_R	Temperature of the reactor, °F
T_C	Minimum temperature of the coolant, °F
t	Elapsed time from the start of the upset to C_i
U	Heat transfer coefficient, BTU/hr-°F-ft²
V	Volume of the reactor, ft³
W	Weight of material (metal, water, reactants), lb
Z	Ratio of catalyst deactivated to CO_2, lb/lb
ΔH_R	Heat of reaction, BTU/lb

11

VERIFICATION OF PROCESS INSTRUMENTATION DATA

11.1 INTRODUCTION

Data verification is a necessity for a successful problem solver. This is true whether the problem solver is working to discover a cure for cancer or working to solve a process plant problem. In all cases, the verification involves use of both human resources and technical resources. While it may be possible to install multiple backup instruments, do frequent instrumentation calibrations, and install elaborate communication devices to avoid the concept of data verification, this level of sophistication is rarely justified in a process plant.

Data verification can take many different approaches. The purpose of this chapter is to elucidate some of the techniques the author has developed and/ or used that are beyond the conventional "check out the instrument" approach. The approaches discussed in this chapter are certainly not all inclusive. They are given for two purposes:

1. To illustrate some of the techniques that are available to use to check out suspect data. As technology growth occurs, newer techniques will cause this list to require continual updating.
2. To serve as an encouragement to the problem solver to take full responsibility for the area of data verification. Often, the problem solver is

Problem Solving for Process Operators and Specialists, First Edition. Joseph M. Bonem.
© 2011 John Wiley & Sons, Inc. Published 2011 by John Wiley & Sons Inc.

tempted to quit and say "I can't do any more until someone fixes the instrumentation."

11.2 DATA VERIFICATION VIA TECHNICAL RESOURCES

The first step in verifying any instrument is a careful review of the instrument specification sheet and a comparison of the specification sheet to the actual field installation. The instrumentation specification sheet will have such information as instrument range, process fluid density, pressure, and temperature. If these values are not correct, or if the range or zero point is set wrong in the field, the data cannot expected to be correct.

A second area to consider when reviewing the instrument specification sheet is actual operating conditions compared to design conditions. That is, how close is the variable being monitored to the design level of flow, pressure, level, and so forth? Many instruments provide a measurement that is based on differential pressure. If the actual measured value is significantly different from the design level, there may be inaccuracy introduced by this deviation. This inaccuracy is often overlooked in the era of digital data acquisition where the measured variable may be shown in 4 or 5 significant figures regardless of what percentage of the range this is. An analog picture that may help in understanding this involves measuring the pressure of an automobile tire. The absolute accuracy of a pressure gage that is calibrated from 0 to 200 psig is much less than that of one calibrated from 0 to 40 psig. An error of 5% of maximum amounts to either 10 psig or 2 psig. This difference in accuracy is quite significant when measuring tire pressures of 30 psig.

This "percent of range" problem is particularly significant for an orifice flow instrument. With an orifice flow meter, the differential pressure across the orifice is proportional to the rate squared. Thus a flow rate of 50% of the maximum design flow rate only provides an output of 25% of the full meter range. The absolute meter accuracy at 25% of range will be much less than that at full meter range.

The concept of reviewing the instrument data sheet first follows the idea of doing the easiest thing first. If errors are found or if the operating conditions are found to be only at a low percentage of design conditions, these can almost always be rectified by simple adjustments of the instrument.

If the review of the instrument specification sheet fails to uncover any explanation for the suspected instrument being wrong, additional steps are required. In a continuous process plant, verification of instrumentation data is complicated by the fact that a shutdown of a process to verify an instrument is prohibitively expensive. Thus data verification must be done through techniques that do not require a shutdown.

Essentially, all instruments consist of both a primary element and a display element. The primary element senses a process variable such as pressure or differential pressure. The display element takes that reading and converts it to the variable being measured such as flow rate, pressure, or level. The primary element is often not accessible without a partial or total plant shutdown. The display element can almost always be tested and replaced or adjusted while the process is in operation. For example, a flow instrument such as an orifice meter both creates and measures pressure drop. Since the orifice plate which creates the pressure drop is installed in the flowing fluid, it cannot be removed from service without at least a partial shutdown. However, the display element can be checked and adjusted or replaced without a shutdown. The emphasis in this chapter is on data verification when the primary element is in question and cannot be removed from service.

Table 11-1 summarizes the typical sources of instrumentation errors of the primary elements. As an example of the concepts in Table 11-1, the reader is referred to Figure 11-1, which shows a typical flow element and potential sources of errors. As shown in Figure 11-1 and Table 11-1, the flow meter can give erroneous results if one of the following occurs:

Table 11-1   Typical sources of primary element errors

Primary Element	Plugging in		Corrosion of Element	Condensation in Tubing
	Element	Tubing		
Flow	X	X	X	X
Level		X		X
Pressure	X	X		X
Differential pressure	X	X		X

X indicates that the primary element could likely be impacted by the condition shown at the top of the table.

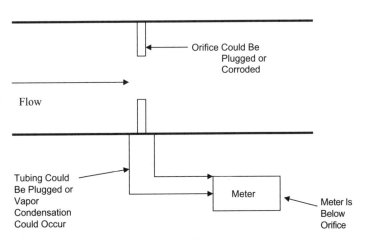

Figure 11-1 Potential flow meter errors.

- *The orifice could be partially plugged.* This would create a greater-than-expected pressure drop for the given flow rate, and would result in the measured flow rate being higher than actual flow.

- *The orifice could be corroded.* This would create a lower-than-expected pressure drop for the given flow rate, since the orifice hole would be bigger than it was in the design. This would result in the measured flow being less than the actual flow. If corrosion of an orifice plate does occur, it might result an orifice hole that is not uniform or is very rough, which might well create greater-than-expected pressure drop and a measured flow being greater than the actual flow.

- *One of the tubing lines leading to the display element could be partially plugged.* This would give a pressure reading at the display element that was different than the actual. The pressure sensors for the meter measure the pressure differential across the orifice. If the high pressure side tubing is plugged, the measured pressure differential will be lower than actual. This will cause the measured flow rate to be less than the actual flow. Conversely, if the low pressure side of the tubing is plugged, the differential pressure will be higher than actual and the measured flow rate will be higher than the actual flow. Possible sources of plugging are solids in the process fluid or freezing of the fluid in the tubing lines.

- *Condensation in the tubing lines can create a false differential pressure.* The comments here assume that the meter is located below the orifice so that condensation will accumulate in the tubing lines. If the meter is above the orifice, any condensation will normally drain back into the process flow by gravity. This condensation can occur when the orifice is in hot vapor service or when the ambient temperature is colder than the temperature of the flowing vapor. If this condensation occurs in the high pressure tubing line and not in the low pressure tubing line, the measured differential pressure will be higher than that created by the pressure drop across the orifice. This will result in the measured flow being higher than the actual flow. The converse is that if the low pressure tubing line has more condensation than the high pressure tubing line, the measured flow will be less than actual flow. If an equal amount of condensation occurs in both the high and low pressure tubing lines and the tubing lines have the same elevation change, then the impact on the measured variable is minimal. Since this is unlikely to occur in practice, the tubing lines are usually insulated and steam traced or sealed with either the process fluid or an instrumentation fluid that is not soluble in the process fluid. A similar situation can occur if the fluid is a liquid that can vaporize in the tubing lines.

- *It is possible for the flow measurement of a vapor stream entraining liquid droplets to experience a similar error.* In this case, the liquid droplets may accumulate in one or both of the tubing lines. While these droplets may vaporize over time, this vaporization will occur at a rate that depends on

Table 11-2 Data verification via technical resources

Type of Instrumentation	Repair Mode[a]	Data Verification Techniques
Flow measurement		
Primary element	Off-line	Insertable flow meters, ultrasonic flow meters, or process analysis
Display element	On-line	
Thermocouples with external wells		
Primary element	On-line	Thermometers, pyrometers
Display element	On-line	
Thermocouples without external wells		
Primary element	Off-line	Infrared pyrometers
Display element	On-line	
Pressure measurement		
Primary element with block valve	On-line	Pressure gage
Display element	On-line	
Level measurement		
Primary element	Both off-line and on-line	Gage glass, Geiger counter or X-Ray
Display element	On-line	

[a] Off-line indicates that the process must be shut down to check or repair. On-line indicates that verification can be made while the process is in operation.

the ambient conditions. In addition, the rate of entrainment may also vary. Thus there is no way to know that the tubing lines are full of liquid or vapor. The preventative steps described above are usually provided by the instrument designer in this case.

A similar description could be provided for each of the types of instrumentation measurements shown in Table 11-1.

Table 11-2 shows possible techniques for validating instrumentation data. The emphasis in this table is verification without requiring a plant downtime. Thus the most obvious approach, replacing an "in-line" instrument, is not covered. The following paragraphs discuss verification techniques for the most common instruments.

11.3 FLOW MEASUREMENT

If a primary flow measurement device (an orifice plate or venturi meter) is suspect and cannot be removed from service, it must be verified by either, or both, a noninvasive external flow measurement device or process analysis.

Table 11-3 External flow measuring devices

Type of Device	Comments
Noninvasive ultrasonic instruments	These can be strapped onto a pipe and used to measure flow.
Transit time	Clean homogenous fluids. Accuracy has been reported from 10% to 70%.
Doppler shift	Used for suspended solids with concentrations from 200 ppm to the percentage level.
Invasive	These require a bleeder and a packing gland arrangement to insert them into the flow.
Pitot tube	Can be highly accurate if sufficient readings are taken across the flow path and integrated to obtain the average flow. Generally for clean fluids only.

The external flow measurement devices are generally not as accurate as the flow measurement devices installed in the process. There are some claims for high accuracy. However, this is usually based on devices that have been calibrated against a known flow rate of the suspect process fluid. If this field calibration against a reliable flow meter is not available, then claims of high accuracy are questionable. In addition, some external measurement devices (pitot tube) require insertion through a bleeder valve with a packing gland arrangement. This packing gland arrangement creates some safety risk, especially in high pressure service. Table 11-3 lists a few of the external devices that are available. This is not meant to be an exhaustive list; in addition, the changing technology world may well allow development of different and/or more accurate devices.

Verification of a flow measurement by process analysis is a relatively simple, but often overlooked, technique. It involves the use of heat and material balances to confirm the validity of a suspect flow meter. It is probably best to consider it as a "one-sided" test. For example, if the heat and/or material balances appear to check, the flow meter is probably correct. If the heat and/or material balances do not check, the suspect flow meter or another flow meter is likely wrong. While these concepts are fundamental to all process engineering, some general guidelines are presented below to aid in the application of these principles.

- *Accumulation may cause a system to appear to be out of material balance.* Thus it may be necessary to expand the traditional concept to utilize the principle of accumulation discussed previously in Chapter 10. That is,

$$AD = I - O - RD \qquad (10\text{-}1)$$

where

AD = rate of accumulation; this could be accumulation of anything: heat or material

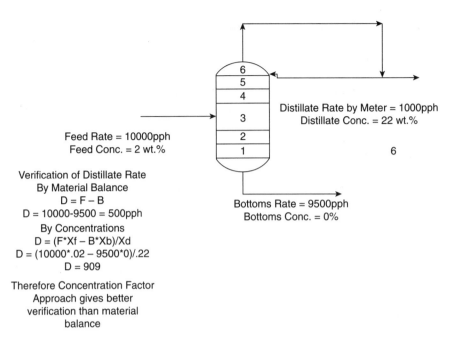

Figure 11-2   Estimating flow by concentration.

I = inflow of material or heat

O = outflow of material or heat

RD = removal of heat or material by reaction

- *Beware of determining flow rate as the difference between two large metered flows.* Using the concept of concentration factor provides more reliable answers. Figure 11-2 illustrates this point. As shown in this figure, if the overhead distillate rate is calculated as the difference between the feed rate and the bottoms rate, a value of 500 lb/hr is obtained, compared to a metered value of 1000 lb/hr. However, if concentrations are used to estimate the overhead rate, a flow rate of 909 lb/hr is calculated.

- *Heat balances can be used to estimate the process or utilities flow to an operation.* For example, the steam rate to a reboiler for a simple fractionation column can be estimated knowing that the heat added by the reboiler and feed must be equal to the heat removed by the process streams and cooling water. Other examples are that the cooling water flow to a tower condenser can be determined knowing the heat removed in the condenser and the inlet and outlet water temperatures, and the "boil up" rate in an evaporator can be determined by the steam rate and the heat of vaporization of the process fluid.

11.4   TEMPERATURE MEASUREMENT

A suspect primary temperature measuring element can often be checked when removed from a thermowell which is installed in the vessel or pipe. However, in some instances, the primary device is not removable. An example of this is the tube metal thermocouples in a furnace. These devices are used to measure the temperature of the furnace tube metal. In order to obtain this measurement, the thermocouples are attached to the tube itself and shielded from the flame, as opposed to being installed in a thermowell. This shielding ensures that the thermocouple does not receive any radiant heat from the furnace flames. A high tube metal reading could be real, or could be associated with a failure of the device or shielding equipment.

In the case of furnace tube thermocouples that are suspect, infrared temperature measurement can be utilized to measure tube metal temperatures through a furnace port hole if it can be accessed safely. Infrared temperature measurement is a noninvasive technique used to determine the temperature of a small area. It can be used to measure reasonably accurately the temperature of a particular spot on a furnace tube. It can also be used to measure the temperature of a flowing solid, or to measure the temperature of a rotating part of machinery.

Infrared temperature measurement requires a great deal of expertise, particularly the measurement of furnace tube temperatures. For example, measurement of furnace tube temperatures may require that the infrared detector be cooled by liquid nitrogen.

11.5   PRESSURE MEASUREMENT

A suspect pressure instrument can, many times, be removed and a gage installed to verify the pressure. However, if there is no way to isolate the pressure instrument or if the pressure tap is plugged, alternative methods of verifying the pressure instrument will be required. Fundamental process analysis can often be used to verify a pressure reading.

The classical chemistry phase rule can often be applied in this situation. This rule is as follows:

$$F = C - P + 2 \qquad\qquad (11\text{-}1)$$

where

F = degrees of freedom (temperature, pressure, and composition)
C = number of components in the mixture
P = number of phases present

Thus for a single component system with two phases (liquid and vapor) present, the degrees of freedom equal one. This means if the temperature is known, the pressure is fixed. In a two-phase binary (two component) system, the degrees of freedom will be two. If the binary system has a fixed composition, fixing the temperature also fixes the pressure since the fixed composition eliminates one degree of freedom. While this may seem to be very basic chemistry/engineering, it is amazing how often this is overlooked in the data verification process. Example Problem 11-1 illustrates how this fundamental approach was used to solve a process problem rather than inventing more complicated theories.

11.6 LEVEL MEASUREMENT

The verification of a level instrument can often be done simply by utilization of a gage glass, if one is available. Other devices such as x-ray and Geiger counters can be used to determine the absolute level based on the fact that liquid is denser than vapor. Both of these techniques work in a similar fashion. A source of the radiation signal is located on one side of the vessel and a detector is located on the other side. The amount of radiation that reaches the detector is inversely proportional to the length of the flow path and the density of the material in the flow path. The material in the flow path includes both the vessel wall and the process fluids. Since the flow path and thickness of the vessel wall are fixed, the only remaining variable is the density of the material in the vessel. If the material is a liquid (higher density), more radiation will be absorbed and the reading on the detector will be lower than it would be if the material was vapor. The radiation absorption of the metal wall can normally be eliminated by a calibration technique. Thus the utilization of radioactive techniques can detect the interface between liquid and vapor. Key factors in the use of these techniques are as follows:

- *Difference in densities between the liquid and vapor.* As the density difference decreases, it is more difficult to measure the level with this technique.
- *Thickness of the vessel wall.* The thicker the vessel wall, the more difficult it is for a signal to penetrate the wall. If most of the radiation absorption occurs in the vessel wall, it will be difficult to utilize this technique to determine the presence of a vapor-liquid interface.
- *Diameter of the vessel.* At a fixed design pressure, larger diameter vessels require thicker walls. These thicker walls will absorb more of the radiation and, as indicated above, the determination of the vapor-liquid interface will be more difficult.
- *Strength of the radiation signal.* The stronger the source of the radiation signal, the easier it is to obtain a strong signal with the detector. However,

stronger signals increase the radiation hazards and the design consideration difficulties.

Again, process analysis can be used to verify the range of the level instrument. In order to verify the absolute accuracy of the level instrument, the zero point and range must be known. The zero point is the level in the vessel that corresponds to a zero reading on the level instrument. This may or may not correspond to an empty vessel condition. The range is the difference (usually in inches) between the zero point and the 100% indication of the vessel level. For example, a 6-ft-high vertical drum with a zero point set at the bottom tangent line of the vessel and a range of 72 in would cover the entire height of the vessel. Thus a reading of 50% on the level instrument would be expected to be 36 in above the bottom tangent line.

If the vessel level instrument is in question, it may be due to an inaccurate zero point or an inaccurate range. The zero point and instrument range can often be checked without taking the vessel out of service. However, some types of level instruments cannot be checked without taking the vessel out of service. In addition, inaccuracies of level instruments can also be caused by internal vessel connection plugging.

If accurate flow meters are available for the inlet and outlet flows, it may be possible to assess the accuracy of the range of the level instrument by accumulating in the drum or removing liquid from the drum and comparing the calculated inventory change from flow meters to the measured inventory change. The approach to be used depends on the orientation of the drum. If the drum is a vertical vessel, the inlet and outlet flows should be set so that there is a significant difference between the two. The measured inventory change should then be compared to the calculated inventory change based on the flow meters. If they compare well, this is a good indication that the range on the instrument is correct. However, the zero point could still be wrong. The zero point can be checked by reducing the inventory in the vessel until there is clear evidence that the drum is empty. The evidence could consist of "blowing through," as evidenced by the outlet flow meter or by a change in temperature which would indicate by the phase rule that there was only a single vapor phase present.

If the level instrument in question is on a horizontal vessel, the same technique can be utilized. However, determining the actual volumetric inventory change will be more difficult due to the curvature of the vessel walls. Tables and techniques are available to readily take this curvature into account. If an accumulation test is run with a horizontal vessel whose level zero point is in error, it is possible to conclude that the level meter is correct due to the curvature of the drum. For example, a change in the level instrument from 65 to 70 (5%) might be confirmed by the difference in flow meters. However, if the actual level were 10% instead of 65%, a 5% level increase would seem in error when compared to the difference in flow meters. To confirm the accuracy of the range of a level instrument on a horizontal vessel, it is best to run two tests at different vessel levels. If both of these tests indicate that, based on differences in flow meters, the level changes appear correct, it can be assumed that the range of the level instrument is correct. The techniques discussed for determining the zero point for a vertical vessel can also be utilized for a horizontal vessel.

11.7 DATA VERIFICATION VIA HUMAN RESOURCES

The role of an operator or mechanic in problem specification was discussed earlier. It was implied that it was imperative that he was cooperative and that all that was necessary was to solicit his data and/or the problem history. Obviously, an uncooperative worker can be a poor source of data by refusing to share his knowledge, presenting erroneous data, or simply acting as a "smoke screen." For example, an instrument technician who was known to have acrophobia may not have checked out an instrument as he claimed, if it required him to be far above the ground.

While this book does not cover all aspects of interpersonal skills, the guidelines below will help the problem solver obtain the maximum assistance from the operating and mechanical workers.

- Respect the person that you are seeking information from as an equal.
- Learn people's names and don't forget them.
- Cultivate relationships by providing answers to questions, explaining your goals, reviewing the results of any tests, and listening to other's thoughts.
- Don't give others the impression that you are too busy to spend time with them.
- Be positive. Assume that people are always trying to do the right things. If a mistake has been made, assume that it was due to inadequate training or done accidentally. Don't take the position that someone did it on purpose. You will be wrong on occasions, but you won't destroy relationships.
- When it is necessary to look over someone's shoulder as they perform their job, try to convince them that they are educating you on their skills.
- Realize that everyone has bad days. If you get an undeserved "chewing out" by an operating or mechanical worker, take your licks and back away from the conversation without becoming defensive. Defending yourself to someone who is irrational is not possible, and you will cut off a source of future data.

11.8 EXAMPLE PROBLEMS

The example problems below illustrate some of the approaches described in this chapter. The first problem illustrates the value of investigating and verifying conflicting data as opposed to simply assuming that one piece of the mysterious data is wrong. The second problem was solved by a careful review of the instrumentation specification sheet. The third problem is a fictitious problem that is included to illustrate how the technique of inventory change can be used to confirm a level instrument.

EXAMPLE PROBLEM 11-1

A vessel in an ethylene refrigeration unit, shown in Figure 11-3, served as a combination surge and knockout drum. Liquid ethylene flowed into the drum and out to various heat exchangers. The mixture of vapor and liquid ethylene leaving the exchangers was returned to the drum where it was separated into vapor and liquid streams. The ethylene refrigeration system was used to cool liquid propane to a temperature of −145°F. It was important to control the liquid level in the drum to ensure that minimal amounts of liquid were entrained with the vapor and to ensure that liquid ethylene was always available for feeding the heat exchangers. The drum operated under a slight vacuum and the normal temperature was −156°F.

Current conditions were an anomaly. The pressure was at normal conditions. This slight vacuum was confirmed by a second, independent pressure instrument. However, the temperature as measured by two independent thermocouples was +10°F. Operating personnel believed that the thermocouples were both wrong. Management had asked that problem-solving resources determine what the real problem was.

The problem solver's approach to using the five-step problem-solving technique was as follows:

Step 1: Verify that the problem actually occurred.

Since there were two independent thermocouples that both indicated a much higher than expected temperature, it was obvious that something was wrong. This was a real problem.

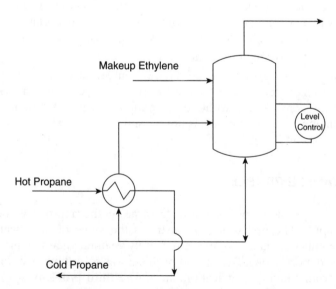

Figure 11-3   Ethylene surge/knockout drum.

Step 2: Write out an accurate statement of what problem you are trying to solve.

The problem statement that was developed was:

> "The temperature on the ethylene knockout drum is reading much higher than normal. While the process appears to be operating normally, this higher temperature may be indicative of a potential problem that will occur in the future. Determine why the temperature on the ethylene knockout drum is reading much higher than the normal temperature of $-156°F$."

Step 3: Develop a theoretically sound working hypothesis that explains as many specifications of the problem as possible.

Using the questions from Chapter 6, the problem solver developed Table 11-4. Based on the above questions, four hypotheses were developed, as follows:

1. There is a leak in the exchanger and the drum contains a mixture of propane and ethylene instead of pure ethylene.
2. The thermowell is so poorly insulated that the temperature instrument is really reading a mixture of the ethylene and the ambient temperature.

Table 11-4   Questions/comments for Problem 11-1

Question	Comment
Are all operating directives and procedures being followed?	All appeared to be correct and being followed.
Are all instruments correct?	The instruments had allegedly been calibrated.
Are laboratory results correct?	Not applicable.
Were there any errors made in the original design?	No. The process was an old one process that was returning to service after repairs.
Were there changes in operating conditions?	No.
Is fluid leakage occurring?	The propane being cooled could leak into the ethylene refrigerant and concentrate in the knockout drum.
Has there been mechanical wear that would explain problem?	The unit had been down for repairs. Repairing of the insulation on the drum might not have been completed.
Is the reaction rate as anticipated?	Not applicable.
Are there adverse reactions occurring?	Not applicable.
Were there errors made in the construction?	See comments above on mechanical wear. In addition, new thermocouples were installed.

3. The new thermocouples were calibrated incorrectly.
4. The level instrument is wrong and there is no liquid level in the drum.

Hypothesis 2 was ruled out by an inspection of the drum that showed the thermowells were well insulated. Hypothesis 3 was ruled out after discussions about the calibration techniques carried out by the instrument technicians.

The problem solver used the phase rule (equation 11-1) shown below to conclude that both hypotheses 1 and 4 were theoretically correct working hypotheses.

$$F = C - P + 2 \tag{11-1}$$

Based on the phase rule, he outlined the following three cases:

1. Under normal situations in the ethylene surge drum there would be a single component ($C=1$), two phases ($P=2$), and the degrees of freedom (F) would be equal to one. That is, the temperature in the surge drum would be fixed by the pressure and should be $-156°F$.
2. If there were two components ($C=2$) in the drum, then the degrees of freedom would increase to two and the temperature and pressure in the drum would be independent.
3. If there was a single component in the drum ($C=1$) but only a single phase ($P=1$), then the degrees of freedom would also be equal to two. Again, the temperature and pressure in the drum would be independent.

However, there are boundaries to the independence of temperature and pressure described in case 2. For example, if there are two phases present and an exchanger leak had occurred (hypothesis 1), the temperature in the drum cannot be warmer than the boiling point of propane. If the liquid level instrument is correct (two phases are present) and, as an extreme case, all of the liquid in the drum is propane, the following analysis could be made:

$C = 1$ (Nothing in the drum, but propane)
$P = 2$ (The level instrument is correct so there must be two phases)

Therefore:

$$F = 1 - 2 + 2 = 1 \tag{11-2}$$

The boiling point of propane at atmospheric pressure is $-44°F$. Thus even if the liquid in the drum were 100% propane, the temperature would be slightly below $-44°F$, not $+10°F$.

As discussed in case 3, if there was no leak in the exchanger, but the level instrument were wrong and the drum was empty (hypothesis 4), the phase rule would indicate the following:

$C = 1$ (Nothing in the drum but ethylene)

$P = 1$ (Nothing in the drum but vapor)

Then:

$$F = 1 - 1 + 2 = 2 \tag{11-3}$$

In this case, the temperature of the ethylene vapor would be limited only by the amount of superheat added by the heat exchangers.

Therefore, the theoretically sound working hypothesis developed was as follows:

> "It is theorized that the mysterious temperatures in the ethylene knockout drum are due to the absence of a liquid level in the drum. Since there is no liquid level in the drum, the vapor can be heated to temperatures above the boiling point of ethylene at the pressure in the drum. In addition, the absence of this boiling liquid results in a very low heat transfer coefficient between the phase in the drum (vapor) and the thermowell. The normal situation in the drum is such that the thermowells are covered with boiling ethylene. The heat transfer coefficient in the normal case is likely as high as $200\,BTU/hr\text{-}°F\text{-}Ft^2$. The heat transfer coefficient from vapor to the thermowell is likely as low as $10\,BTU/hr\text{-}°F\text{-}Ft^2$. With this exceptionally low coefficient, the heat gained from the atmosphere flowing through the insulation and heating the thermocouple may create an additional error source."

Step 4: Provide a mechanism to test the hypothesis.

Although the hypothesis could have been checked by simply raising the liquid level in the drum, this was deemed too risky. Since the drum served as a knockout drum, there was a risk that a high level would cause liquid ethylene to be carried over into the compressors. The problem solver requested that the instrument technician (who had recently checked the level instrument and determined that the range and zero point were as they should be) accompany him to again check the instrument. He explained to the instrument technician why he thought that the level instrument was incorrect. When they jointly inspected the installation, they found that the steam tracing that was used to ensure that there was no liquid accumulation in the tubing lines was turned off. When this tracing was returned to service, the level instrument began to function correctly.

Step 5: Recommend remedial action to eliminate the problem without creating another problem.

In this case, the mechanism to test the hypothesis was the remedial action. In order to prevent future occurrences of this problem, a statement was added to the startup procedure that called for the operators to check that the steam tracing had been put into service.

Lessons Learned While it is true that an experienced engineer would not need to use the phase rule to analyze this problem, that is only because he knows by experience and intuition how to analyze similar situations. In applying his experience and intuition, he is using the phase rule whether he knows it or not. This problem might be considered too simple for a detailed analysis; however, it does illustrate two important points. In the first place, there is great value in doing a fundamental analysis instead of making the assumption that the first analysis (the thermocouples are both wrong) is correct. It would be unlikely that both thermocouples were incorrect unless there was an external circumstance such as if they were incorrectly calibrated or the heat transfer to the thermocouple was not what it should have been. As indicated, this could be due to excessive heat loss from the thermowell due to poor insulation or poor heat transfer (caused by the absence of liquid) from the vessel contents to the thermowell. The application of the phase rule was very helpful in indicating that the level instrument which had been checked out was truly in error.

The problem also indicates that it is often necessary to get an operator or mechanical personnel to recheck something that has already allegedly been checked. This requires tactfulness and a good working relationship. This good working relationship is developed by the principles indicated in Section 11.7. In this example problem, it was enhanced by the problem solver going with the instrument technician to look at the level instrument for a second time.

EXAMPLE PROBLEM 11-2

A new polymer was to be produced in a test run in a commercial polypropylene plant. This new polymer was to be a copolymer produced by using ethylene as a comonomer in reactors normally polymerizing only propylene. Careful preparations were made for the test run, including reviews of batch and pilot plant operations producing the copolymer. This data indicated that the ethylene conversion should be about 85% during the planned operation. The test run plans called for careful monitoring of all variables, especially the ethylene conversion. The ethylene conversion was to be monitored using the propylene and ethylene flow meters, the ethylene content of the polymer, and the ethylene content of the unreacted gases. Because of the reliance on feed flow meters, these were carefully checked out by the instrument technicians prior to beginning the operation.

After the test run was started and steady state was established, the calculated ethylene conversion based on flow meters and the ethylene content of the polymer was essentially 100%. This was in contrast to the ethylene content of the unreacted gases, which indicated that the ethylene conversion was significantly less than 100%. While the unit ran well, the problem solver was greatly concerned about the deviation from the anticipated conversion based on calculations made using the feed meters. He convinced management that even though this discrepancy between the anticipated conversion and the apparent actual conversion did not seem to be a problem on this run, that it might be a problem on future runs. They agreed that he should spend some time investigating the problem.

The problem solver's approach to using the five-step problem-solving technique was as follows:

Step 1: Verify that the problem actually occurred.

Since the test run had been carefully planned and the instruments had been checked out ahead of time, there was no doubt that there was a problem. There was some doubt about whether the problem merited attention. The problem statement developed by the problem solver had to address the incentive for working the problem.

Step 2: Write out an accurate statement of what problem you are trying to solve.

The problem statement that was developed was:

"The calculated ethylene conversion based on the ethylene feed flow meters and the ethylene content of the polymer is about 100%. This is in conflict with the pilot plant data and with the analysis of the unreacted gases, which shows the presence of ethylene. If the flow meters are correct, then either the ethylene content of the polymer or the ethylene content of the unreacted gases is wrong. The process control strategy requires that all three variables (feed flow meters, ethylene content of the polymer, and ethylene content of the unreacted gases) be measured correctly. Determine why the ethylene conversion is significantly higher than anticipated based on feed flow meters."

Step 3: Develop a theoretically sound working hypothesis that explains as many specifications of the problem as possible.

Using the questions from Chapter 6, the problem solver developed Table 11-5. Based on the questions in the table, four hypotheses were developed as follows:

1. There was a design mistake in sizing the ethylene flow meters.
2. There was an error made in the installation of the ethylene flow meters.
3. The unreacted gas analyzer had been incorrectly calibrated for ethylene, or else the installation or design of the analyzer made it unacceptable for ethylene monitoring.
4. The technique for analyzing for ethylene in the polymer was either incorrect or being done incorrectly.

Table 11-5 Questions/comments for Problem 11-2

Question	Comment
Are all operating directives and procedures being followed?	All appeared to be correct and being followed.
Are all instruments correct?	The instruments had allegedly been calibrated. The ethylene flow meters and the analyzer for measuring the ethylene content of the unreacted gases were being used for the first time.
Are laboratory results correct?	The technique for measuring the ethylene content of the polymer had been closely cross checked against known standards.
Were there any errors made in original design?	Since this was the first time that the ethylene meters and analyzers had been used, their design could be in error.
Were there changes in operating conditions?	Not applicable.
Is fluid leakage occurring?	Not applicable.
Has there been mechanical wear that would explain problem?	Not applicable.
Is the reaction rate as anticipated?	Higher-than-anticipated ethylene conversion might explain the apparent discrepancies.
Are there adverse reactions occurring?	Adverse reactions are highly unlikely.
Were there errors made in the construction?	Since both the ethylene flow meters and the ethylene analyzer were being used for the first time, this had to be considered.

While any of the four hypotheses could be selected as a "theoretically sound working hypothesis," the problem solver elected to do some further investigation prior to proposing a working hypothesis. Since the easiest and quickest thing to do was to review the instrument specification sheets, he elected this route. In addition, he recognized that the pretest instrument check by the instrument technicians did not involve a review of the instrument specification sheet. A review of the specification sheet indicated the following:

Table 11-6 Specification and actual data

Variable	Specification Sheet	Actual Data
Pressure, psig	735	735
Temperature, °F	90	75
Critical pressure, psig	735	735
Critical temperature, °F	50	50
Compressibility (Z)	1.0	0.6
Density, lb/ft^3	3.56	6.10

The original instrumentation design used a density that was significantly less than the actual density because the gas compressibility that occurs at high pressures was not considered. In theory, the influence of pressure on gas density should be adequately calculated based on a ratio of actual pressure to atmospheric pressure. However, at high pressures, the gas is more compressible than would be indicated by this ratio alone. Thus a compressibility factor must be included. This compressibility factor is available in tables and handbooks. This factor was not considered in the instrument design. There was also a difference in actual and design temperatures, but this was not significant. These calculations led to the following theoretically sound working hypothesis:

"It is theorized that the discrepancies between the actual ethylene conversions and the indicated ethylene conversions are due to the fact that the density for the ethylene flow meters was calculated assuming that the compressibility was 1. This resulted in a calculated density that is 40% below actual density. Since the measured flow rate is directly proportional to the square root of the density, this error results in a measured flow rate that is 30% below the actual flow rate. This single error explains the discrepancies observed."

Step 4: Provide a mechanism to test the hypothesis.

The mechanism for testing the hypothesis was very simple. Since the error was in the calculations, it was easily corrected by changing the range on the flow instrument to correct for the incorrect density. After this correction was made, the flow rates, ethylene content of the unreacted gases, and ethylene content of the polymer were all consistent with the anticipated results.

Step 5: Recommend remedial action to eliminate the problem without creating another problem.

In this case, the testing mechanism was the permanent solution. It should be noted that in this problem, there was a temptation to simply modify the process control computer with a "fudge factor" to adjust for the error. This was easy to do, but it brought with it the risk that at some future time, the factor that was used in the computer would be removed because someone did not understand what it was. It was much more fool proof to change the instrument range and revise the instrument specification sheet.

Lessons Learned   It should be noted that the error in the instrument specification sheet was only discovered after the problem solver realized that the pretest instrument checkout done by the instrument technicians only included a physical check of the instruments against the range given on the instrument specification sheet. They had no way of knowing that the density used in calculating this range was incorrect. The problem-solving lesson to be learned is that the problem solver needs to know the boundaries of the mechanics or operators when they are asked to check out a piece of equipment. If they are starting with an incorrect calculated setting, they can only confirm that the instrument or piece of equipment is set to that value.

While it could be argued that the effort taken to resolve this discrepancy was not justified, the discrepancy was a particular concern since the flow meters were a key to changing grades. In addition, some grades produced at high ethylene contents could cause process shutdowns if control were lost. As a general rule, discrepancies in data or things that we do not understand will almost always cause future problems.

This example problem also illustrates again the value of doing the easiest calculation or data review first. While it seemed hard to believe that the design team would ignore the compressible nature of ethylene at high pressures, a simple review of the instrument specification sheet showed that this in deed was the case. Other hypotheses described earlier would have been more involved and taken longer to pursue.

EXAMPLE PROBLEM 11-3

This problem is included to show the calculation technique discussed earlier. The 5-step problem-solving approach is not considered. In addition, the tables that show liquid volume as a function of level in a horizontal drum are not included. For simplicity, the head volume of the drum is not included. In order to illustrate the calculation technique, it is assumed that the bottom connection of the differential pressure type level indicator is partially plugged so that the indicated level is 15% greater than the actual level.

A pump taking suction from a horizontal vessel continued to experience intermittent periods of operating below the pump curve. At times, it operated well. After an extensive problem-solving analysis, it appeared that the level instrument was incorrect. The level instrument was checked by instrument technicians and the range and zero point appeared to be correct. The problem solver decided that the only way to confirm the accuracy of the level instrument was with a plant test using the inlet and outlet flow meters. He planned to use the flow meters to calculate the level change in the drum. He would then compare this change to the measured change to try to determine if the level instrument was correct.

A summary of the test is shown in Tables 11-7 and 11-8.

Table 11-7 Summary of test basis

Drum size	
Diameter, ft	10
Length, ft	30
Fluid density, lb/ft^3	30.6
Inlet flow – outlet flow, lb/hr	10000
Test time, min	30
Volumetric flow, ft^3	163.3

Table 11-8 Test run results

	Test 1	Test 2
At Test Start		
Indicated level, %[a]	25	65
Actual level, %	10	50
Actual contained volume, ft^3	123	1178
At Test Completion		
Actual contained volume, ft^3	286	1341
Actual level, %	17.9	55.5
Actual level change, %	7.9	5.5
Measured level change, %	6.1	5.8
Error, %	23	5

[a]The problem description indicated that the level instrument showed 15% higher than the actual level.

An examination of Table 11-8 indicates the risk of running an accumulation test at only one level. If the test is run at a starting indicated level of 65%, it would be easy to conclude that the actual level change and calculated level change were very close. However, if the test run is started at an indicated level of 25%, the conclusion would be that there is a significant difference between the actual level change and calculated level change.

NOMENCLATURE

F Degrees of freedom (temperature, pressure, and composition)
C Number of components in the mixture
P Number of phases present

12

SUCCESSFUL
PLANT TESTS

12.1 INTRODUCTION

Step 4 of the five-step problem-solving procedure often involves a plant test of some kind. These tests can vary from very straightforward to very complicated. While some of these comments were included in Chapter 3, conducting a successful plant test is so important to solving many problems that it deserves a separate chapter.

The reader of this chapter may or may not be involved in planning a plant test. However, it is likely that he will be involved in planning, reviewing, or executing of the test. Thus it is important that he understand the criteria and the elements involved in a successful plant test.

As indicated earlier, a successful plant test is one that either confirms or disproves the hypothesis. An unsuccessful plant test is one that fails to either confirm or negate the hypothesis. In addition, it is important that the plant test be conducted in such a fashion that it does not create a major problem for plant operations. The concept of analyzing problem solutions for potential problems was discussed earlier. Before conducting any plant test, a well-thought-out potential problem analysis should be completed.

The Russian nuclear power plant disaster at Chernobyl was at least partially due to failure to complete a well-thought-out potential problem analysis ahead of the test. The test was designed to determine whether, in the event of a

Problem Solving for Process Operators and Specialists, First Edition. Joseph M. Bonem.
© 2011 John Wiley & Sons, Inc. Published 2011 by John Wiley & Sons Inc.

reactor shutdown, enough power was available to operate the emergency equipment and core cooling pumps until the diesel power supply came online. It was to be conducted during the planned shutdown of reactor 4. The following contributed to the disaster:

- Only six to eight control rods were used during the test, despite there being a standard order stating that a minimum of 30 rods were required to maintain control.
- The reactor's emergency cooling system was disabled.
- The test was carried out without a proper exchange of information between the team in charge of the test and personnel responsible for the operation of the nuclear reactor.

12.2   INGREDIENTS FOR SUCCESSFUL PLANT TESTS

All successful plant tests will require the following:

- A full evaluation of instruments and laboratory procedures to be used for the test.
- A careful statement of what results are anticipated and how the anticipated results will be evaluated. This will almost always involve a significant amount of pretest calculations.
- A complete and well-thought-out potential problem analysis including "trigger points" which, if violated, will cause the test to be terminated.
- A careful and detailed explanation to operating personnel about the test.
- A formal post-test evaluation and documentation.

These items are discussed in more detail in the following paragraphs.

12.3   PRETEST INSTRUMENT AND LABORATORY PROCEDURE EVALUATION

One might think an evaluation of instrumentation prior to a plant test would involve only instrument technicians "zeroing" and calibrating instruments. In addition, however, the instrument specification sheets themselves should also be reviewed if there is any doubt at all about the instrument validity. For example, as discussed in Chapter 11, during a plant test in a high pressure process, it can be difficult to get a good material balance using instrumentation. All instrumentation was checked and confirmed to be as indicated on the specification sheets. On a closer evaluation of the specification sheets, it was discovered that no allowance had been made for gas compressibility at the high pressures. This would result in a higher density than that calculated using the ideal gas relationships. There will probably be value in reviewing the specification sheets of all new instruments to be used in a plant test. Depending on the complexity of the instrument specification sheets, this review may require the assistance of a process or instrument engineer.

While laboratory procedures may be thought of as the domain of the chemists, if results do not appear to make sense, the problem solver should not assume that the procedure is correct or is being correctly followed. As an example, during a startup of a chloride removal system, strange laboratory results were reported. The procedure developed to test for the level of the chloride ion called for acidification (reducing the pH into the acid range) of the sample prior to running the test for chlorides. The laboratory technician used the most available acid (hydrochloric acid) for the acidification step. This was discovered only when a tactful process engineer asked if he could watch the laboratory technician perform the test. On careful examination of the procedure, a footnote was found that indicated that nitric acid should be used in the acidification step.

These two examples are provided to indicate how the process engineer or problem solver must be involved in every step of the problem-solving efforts and/or plant test preparations, regardless of which discipline has primary responsibility. In these two cases, a more thorough review of procedures prior to the plant tests would have allowed the problem solver to discover that the incorrect gas compressibility was used as the basis for flow meter calculations and that the laboratory procedure was not as clear as it should have been. That is, the procedure calling for the use of nitric acid should not have been in a footnote.

12.4   STATEMENT OF ANTICIPATED RESULTS

Essentially, all plant tests are directed toward improving plant operations. However, in order to obtain an adequate evaluation of the plant test, it is imperative that the anticipated results be stated as quantitatively as possible. In addition, the variable or variables to be used to monitor the results of the plant test should be quantified. Both the positive and potential negative results should be spelled out. For example, a plant test on operating a fractionation column at a 15% increase in reflux rate to improve the purity of the distillate might have anticipated results stated as:

> "It is expected that this plant test will provide an increase in distillate purity of 0.5 weight percent. The distillate purity will be measured by the standard laboratory test procedure which is accurate to within 0.1 weight percent. An increase in tower pressure drop of 0.5 psi, with no indication of tower flooding, is anticipated. Tower flooding will be monitored by deviations from anticipated values of tray efficiency, tower pressure drop and heat balance closure."

There are several important aspects to notice in this statement. The fact that these are anticipated results means that calculations have been performed to determine the impact of the increase in reflux rate. For these calculations to be meaningful, similar calculations must be performed on the "base case." That is, fractionation calculations should be performed for the normal operating conditions. These fractionation calculations are based on theoretical trays. The base case tray efficiency can be determined by varying the number of theoretical trays in the calculations until the calculated compositions match the laboratory results. At this point, the tray efficiency can be determined by dividing the number of theoretical trays by the actual number of trays. This same tray efficiency can be utilized to estimate the change in distillate purity with the increased reflux rate.

If the anticipated distillate purity is not obtained, it is an indication that the tray efficiency has decreased. This could be the result of some kind of tray overloading condition. In a similar fashion, the tower heat balance closure can be determined for the base case. Heat balance closure is the difference between the heat added, including all heat sources (feeds and external heat source), and the heat removed with the products and cooling sources. In theory, this value should be zero. However, it is rarely zero, due to meter errors and heat losses. If the heat balance closure becomes worse as the reflux is increased, it is likely due to tower flooding. Tower flooding can cause liquid to be entrained from the top of the tower into the reflux accumulator. The heat balance calculations assume that anything that enters the reflux accumulator comes out of the tower as vapor. If some of the material going to the reflux accumulator is, in fact, liquid entrainment, the heat balance calculations will show more heat being removed from the condenser than is the actual case. This will result in a change in heat balance closure; this was illustrated in Chapter 8.

The fact that an anticipated tower pressure drop increase of 0.5 psi is indicated means that tray pressure drops have been calculated. The value in doing tray pressure drop calculations is twofold. It is necessary in order to assess whether tray flooding at the higher reflux rates would be anticipated. In addition, it provides an anticipated pressure drop which can serve as a trigger point to abort the test if exceeded.

While it is not anticipated that a process operator would have capability to perform these calculations, the calculations need to be done in order to set the basis for the plant test. These calculations can best be made by a graduate chemical engineer. The problem solver, whether he is an engineer or operator, needs to be aware of the need for these pretest calculations.

In addition to the pretest calculations, the anticipated results include a statement about the accuracy of the laboratory test. A plant test where the anticipated distillate purity increase is 0.5 weight percent would be meaningless if the laboratory accuracy was only ±0.5 weight percent.

The example of a fractionation column is very straightforward and easy to quantify with both statements and calculations. A more difficult plant test might be one in which a new catalyst with anticipated higher reactivity was to be tested. A similar technique to that described Chapter 9 could be utilized to determine the actual higher reactivity of the new catalyst. The simplified kinetic constant of the base case catalyst could be determined knowing the reactor residence time and the reactant concentrations. The simplified kinetic constant for the new catalyst could be determined using the same variables. A comparison of these two kinetic constants could be made to determine the increased reactivity of the new catalyst. While this is more involved than the less rigorous method of just comparing catalyst efficiencies (pounds of product produced per pound of catalyst) it avoids the need for the plant test to be run at the exact same residence time and reactant concentration. Plant tests of new, high-reactivity catalysts are often conducted at lower feed rate, production rate, and/or reactant concentration to avoid potential problems that might be associated with the higher reactivity. Catalyst efficiency will be impacted by these changes. Thus a comparison based only on catalyst efficiency will not be valid. This type of more complex plant test often involves considerations that, on the surface, seem difficult to assess. In many situations, it is possible to assess these considerations using fundamental chemical engineering skills. Two example problems are discussed later.

12.5 POTENTIAL PROBLEM ANALYSIS

A complete, well-thought-out potential problem analysis is mandatory. A trigger point should be developed for each major potential problem. If these trigger points are violated, the test will be terminated. Most of the variables considered in the statement of anticipated results will be considered in the potential problem analysis. While the concept of potential problem analysis is closely related to the previous discussion, there are some additional considerations. Some of these are related to safety. While the list below is not an all inclusive list, it is a list of the types of safety-related questions that should be considered:

1. Are any new chemicals being used?
2. Have byproduct reactions and byproducts been evaluated?
3. Are any new chemicals or reaction byproducts compatible with the existing materials of construction?
4. Are the test operating conditions outside acceptable ranges?
5. Are there any proposed conditions that will cause the safety release systems to be inadequate? For example, a higher reactivity catalyst might cause safety release facilities to be undersized.
6. Will operating condition changes cause a highly reactive chemical to concentrate to an unsafe level or be at unsafe conditions? Some examples are:
   - Oxygen concentrating in a vent to the point that the oxygen-hydrocarbon mixture is in the explosive range.

- Operating conditions for systems handling ethylene or acetylene reaching a pressure and/or temperature that can result in thermal decomposition.
- Operating temperatures for a thermally sensitive polymer being so high that decomposition, with release of highly reactive monomers, occurs.

7. Are there changes in operating conditions which seem innocuous, but could, along with a single unexpected occurrence such as a utility or mechanical failure, lead to catastrophic results?

The above list, as indicated, is not inclusive, but only serves as example of the types of safety items that should be considered. Any list of this type must not be considered a check list. It should be considered as a guide line only. Check lists often have a way of defeating their purpose by allowing the person responsible to simply check off items. In the pretest work for a plant test, serious consideration and significant calculations should be completed to ensure that no safety-related problems will occur.

In addition to safety-related items, an analysis should be made of what kind of things could go wrong (potential problems) during the plant test. As indicated earlier, each potential problem should have a trigger point. If this trigger point is violated, the plant test will be terminated. The value of having pretest trigger points is that they can be calculation- or logic-based when time is available for careful planning. This is opposed to waiting until unexpected events occur during the plant test and intuition becomes the mode of decision making. Examples of these trigger points are:

1. A new catalyst introduced into the reactor might lead to fouling of the heat transfer surface. In this example, the trigger point should be heat transfer coefficient. A trigger point of reactor temperature only will not be sufficient to determine if the heat transfer surface is fouling. The trigger point for the heat transfer coefficient should be set high enough to avoid the possibility that a small increase in catalyst rate will cause an uncontrolled increase in reactor temperature. Chapter 4 discussed a real-life example of this type of event.

2. A change in the reflux rate to a distillation column that should lead to an increased purity should include a trigger point to allow monitoring of unexpected tray performance deterioration. One possibility would be a trigger point specification on column pressure drop. The anticipated affect on column pressure drop could be calculated prior to the test and monitored during the test.

3. A plant test on adding a reagent to an exothermic batch reaction at a rapid rate might be monitored by a trigger point of temperature increase in the initial 2 or 3 min of reagent addition. Calculations prior to the test could determine the maximum temperature increase required to avoid

exceeding the maximum desired reactor temperature. If this calculated initial value was exceeded, the test would be terminated. Chapter 10 includes an example of this type of trigger point monitoring.

Hypothesis testing in a commercial plant may involve using results based on laboratory or pilot plant studies. Sometimes these results are based on equipment that does not simulate the commercial process exactly. A potential problem analysis should include possibilities associated with differences in equipment such as:

1. Did the reactor used in the studies simulate the commercial facilities? Considerations must be given to:
   - Simulation of a continuous reactor with a batch reactor.
   - Differences in mixing patterns and regimes in a laboratory or pilot plant reactor and a commercial reactor.
   - The reduced heat removal capability in the commercial reactor caused by the smaller heat transfer area to reactor volume (A/V ratio). Scaleup from pilot plant facilities to commercial facilities almost always involves a reduction in the A/V ratio, making heat removal more difficult.
2. Did the equipment used in the laboratory test adequately simulate commercial equipment? For example, chemical compound stability studies are often determined in a laboratory oven. If the material will be subject to high shear rates in the commercial equipment, laboratory oven studies may not be sufficient.

12.6   PLANT SPECIFIC REQUIREMENTS

The series of questions and comments listed above is not meant to supplant any specific company or operating plant requirements. Such steps as safe operating committee reviews, management of change reviews, OSHA-mandated reviews and/or peer reviews will still be required.

12.7   EXPLANATION TO OPERATING PERSONNEL

An explanation of the plant test should be given to operating personnel (and perhaps mechanical personnel). This explanation can be conducted in a training session, with a written handout, or through one-on-one discussions. The following items should be included:

1. *Purpose of the test*: The value of the test for the overall company objectives should be explained to the operating and mechanical personnel. The

plant test could be directed at solving an operating problem that will make for an easier job, or producing a superior product that will lead to increased sales. Regardless of the goal of the test, this is an opportunity to build enthusiasm for the test by explaining that the success of the enterprise leads to both job security and promotional growth for the individual.

2. *How safety was evaluated*: Unfortunately, most hourly personnel have had experiences with plant tests or plant changes where safety was not adequately considered. This explanation should consider new chemicals, new laboratory procedures, and new operating or mechanical conditions/ procedures. Many operating companies have safety and peer review committees that review new operating changes. The operations and mechanical workers should be informed of the details of these reviews. In addition, if new chemicals are involved, the MSDS sheets should be reviewed with the appropriate personnel and posted in visible places.

3. *Why the test will work*: This will be an opportunity to explain to the hourly worker the theory behind the proposed test. Most theoretical explanations can be expressed in terms that even people without engineering degrees can understand. This is an important step and should not be considered just a requirement, but should be looked on as an opportunity to educate and obtain "buy-in" for the test.

4. *How the test will be evaluated*: The purpose of this explanation is twofold. It is an opportunity to explain exactly what will be considered in evaluating the test. It may be desirable, as part of the test, to have the hourly worker fill out data sheets. The accuracy of this will be a function of how well he understands how the values will be used. In addition, this will be a time to explain that a test that does not prove the hypothesis is not a failed test. The only failed tests are those that are not conclusive.

5. *If the test works, what he gets from it*: The hourly worker will often have questions such as:
   - If the test is successful and changes in operating or mechanical procedures are required, will it make my job easier or harder?
   - Will these changes reduce manning or limit future addition of jobs?
   - If the test is successful, how do I get any credit for doing all of this extra work?

These are questions that are often not brought up. However, they are almost always in the minds of the hourly workers. Ignoring them will usually leave it to the operator or mechanic to assume the worst case answers.

12.8 FORMAL POST TEST EVALUATION AND DOCUMENTATION

The formal post-test evaluation and documentation phase of the plant test is often one of the most overlooked areas of plant test execution. The desire to improve organization efficiency often creates pressures to move on to the next plant test involved with the current problem or to move on to the next problem to be solved. Whether the test proved or disproved the working hypothesis, the benefits of the documentation are to provide a lasting reference that will help avoid repeating the test or, even worse, avoid future changes that will cause the problem to recur. In addition, as indicated earlier, the writing process will often clarify the thinking process and improve problem-solving activity.

A plant test that disproves the hypothesis is often not documented because the problem solver does not see this as important. He feels the need to move on and work on the next hypothesis to get to a problem solution. The failure to document successful plant tests that disprove the hypothesis will often cause a similar hypothesis to be proposed at some future time. Thus the failure to document the work, because of the alleged need to be efficient, will cause a loss in organization efficiency. A comment often made by many experienced operating and/or mechanical personnel when presented with a proposed plant test is, "We tried that before and it didn't work." This is an indication that formal documentation of previous tests is not being done.

Even more serious is the failure to document a plant test that proved a working hypothesis. In this scenario, changes are made which eliminate the problem. Several years later, a proposal is made to reverse the changes to try to solve another problem, increase production, or improve product quality. When the question "Why are we operating at these particular conditions?" is raised, no documentation exists to show the fact that changes were made to eliminate an operating problem. Thus the assumption is made that the conditions can be returned to the previous conditions.

Another scenario is the case of a failed plant test. That is one that neither proves or disproves the working hypothesis. Rather than making another attempt with an improved plant test that will be successful (prove or disprove the hypothesis), the approach is just completely abandoned. No documentation is done to indicate why the test was not conclusive. At a later point in time, the hypothesis is reintroduced. When a similar test is proposed, the memories of both technical and operating personnel are that "We tried it before and it didn't work." Actually, the validity of the hypothesis is unknown. What is known is that the test did not prove that the hypothesis was correct. The converse, that the test proved that the hypothesis was incorrect, is not true. Because the recollection is that the test did not work, the hypothesis is abandoned. A good documentation of the previous test would have indicated that what failed was the test itself, rather than the correctness of the hypothesis.

The size of the actual document should be minimized. It should include items such as the objective of the test, the test procedure, the test results, and the conclusions. In addition, any comments concerning safety should be included. Because of the technical conclusions and the possible need to incorporate changes into operating conditions or procedures, the final document should be approved by both operating and technical management.

12.9 EXAMPLES OF PLANT TESTS

Some plant tests to verify hypotheses will be simple and some will be more complicated. The next few pages present two examples of involved plant tests that were directed at improving a product and/or process. While they are not directly related to solving plant problems, they are real world examples that illustrate the concepts discussed above. These same concepts are applicable to plant tests directly associated with solving plant operating problems.

EXAMPLE PLANT TEST 12-1

A synthetic rubber producer was pressured by customers to change their stabilizer to one that would continue providing a product that did not discolor due to stabilizer oxidation, but would have better product-related properties. Based on the customer request, a new stabilizer was identified by the technology organization. Tests to confirm the stabilizing and nondiscoloring properties were conducted at different temperatures in laboratory ovens. These tests were conducted by first mixing the stabilizer with the polymer at ambient temperatures in low-intensity mixers. The polymer containing the stabilizer was then heated and held at elevated temperatures in the oven. These tests confirmed that the product did not discolor and that the molecular weight of the polymer did not change even after several hours in an oven. Thus the new stabilizer seemed to meet the goals, that is, the product did not discolor and the molecular weight did not decrease, indicating that the new additive was an effective stabilizer.

The actual synthetic rubber process consisted of a polymerization section where the stabilizer was added to a slurry of rubber and water. This slurry was then pumped to the finishing section where the rubber was dried to remove water, extruded, and rolled on hot mills to put the final product into a form that could be boxed and shipped to the customer.

A plant test was scheduled to assess the utilization of this new stabilizer in the plant. The hypothesis being tested was that the new stabilizer could be added effectively in the plant and that it would be as efficient as indicated in the laboratory results. Since the laboratory data indicated that the stabilizer was very effective, only some of the steps listed above were implemented. These steps and what was actually done are summarized below:

1. *Pretest Instrument and Laboratory Procedure Evaluation*: For the most part, this was done in a satisfactory manner. The laboratory was prepared to analyze the stabilizer and, in addition, instruments were adjusted for the slightly different density.
2. *Statement of Anticipated Results*: Since the technology experiments indicated that such good results were obtained, very little was done to

prepare a statement of anticipated results. It was just assumed that the stabilization properties would be comparable to those of the existing stabilizer, or even better. That is, that there would be no change in molecular weight during the finishing operation.

3. *Potential Problem Analysis*: Actually, very little was done to develop a potential problem analysis. The technology work was so convincing that this new stabilizer would meet the criteria of not discoloring and avoiding molecular weight change in the finishing operation that this phase was just ignored. It was anticipated to be a "boring test."

4. *Explanation to Operating Personnel*: This was not done.

5. *Formal Post-test Evaluation and Documentation*: As discussed below, the test disproved the hypothesis so conclusively that it was mandatory to document the results of the test.

Since, based on technology testing, it appeared that the new stabilizer had stabilizing properties equivalent to those of the current stabilizer, all operating conditions such as polymer molecular weight and amount of stabilizer added were held constant and operations were simply switched over to the new material. Very soon, the molecular weight of the product leaving the finishing operations began to decrease rapidly, even though the molecular weight in the polymerization operations did not change. This was completely the opposite of what would be expected if the stabilizer was performing as anticipated based on the laboratory tests. The lack of a well-thought-out potential problem analysis before the test was now causing a panicked problem-solving attempt during the test. The two different groups involved (operations and technical) began pursuing different approaches.

The operating group increased the molecular weight in the polymerization section in an attempt to make on-specification product leaving the finishing section. Since the molecular weight in the polymerization process was inversely proportional to the monomer conversion, it was necessary to reduce the monomer conversion in order to increase the molecular weight. This lowered monomer conversion tended to overload the monomer recycle system. The operations continued to deteriorate. The molecular weight of the finished product continued to fall and the monomer conversion continued to be decreased as a compensatory action. The test was aborted after only a few hours of operation.

The technical/technology group initiated problem-solving activities. An analysis of the final product indicated that there was very little stabilizer present. This brought up several potential hypotheses:

- An incorrect amount of stabilizer was being added.
- The stabilizer was being washed off the polymer in the water slurry operation.
- The stabilizer was being vaporized in the finishing operation, even though long-term technology tests in the ovens indicated this would not occur.

- The laboratory results were wrong and stabilizer really was present at the desired concentration.
- There was the target amount of stabilizer in the polymer; it was just not an effective stabilizer. This would be in conflict with the technology data.

An evaluation of the data after the test was aborted indicated that the stabilizer was being added at a correct rate and was not being washed off the polymer in the water slurry operation. However, it was being vaporized in the high shear zone of the extruder and hot mills. These zones were not only at elevated temperatures, but the high shear caused large amounts of the surface to be exposed to the atmosphere. This was a radically different condition than what was experienced in the low-shear environment of the ovens.

Lessons Learned   There are several lessons that can be learned from this plant test. They involve areas such as the failure to conduct a potential problem analysis, the failure to provide an explanation to operating personnel, the blind acceptance of technology data without questioning the basis for conclusions, and the attitude that this is going to be a "boring plant test."

A potential problem analysis would have very likely uncovered that the shear rates in the extruder/mills were vastly different than those in the very low-shear drying ovens. This consideration would have generated additional technology work to try to simulate the high shear rate. This additional work would have likely shown that large amounts of stabilizer were lost in the high-shear-rate studies. If the proposed stabilizer still appeared to be the best available option, the plant test could have been conducted at much higher stabilizer addition rates to compensate for the vaporization of the stabilizer in the extruders and mills. A potential problem analysis would have also led to consideration of the overloading of the monomer recycle system that occurred when the conversion had to be dropped to a low level to obtain the specification molecular weight in the finished product. Plant tests where technology data is accepted blindly and/or where the test is considered to be boring will almost always result in failure of some kind.

The lack of a careful and detailed explanation of the plant test to the operating personnel did not impact the results of this first test. However, when another test was scheduled, there was a considerable amount of tension caused by the failure to adequately explain the goals and purposes of the first test.

EXAMPLE PLANT TEST 12-2

As a second illustration of a plant test, consider the following example. It was desirable to test the capacity of a single reactor in a plant that used two reactors in series. This would allow assessment of the possible, low-cost debottlenecking of the existing plant as opposed to the alternative of building a new plant. If the single reactor could be operated at the same capacity as the exist-

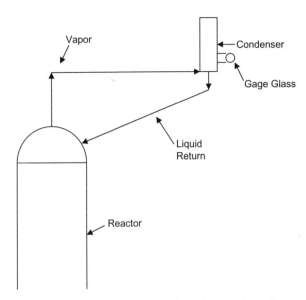

Figure 12-1   Reactor overhead condenser schematic.

ing two reactors in series, it would be possible to increase the existing plant capacity by at least 50%. The hypothesis to be tested was that a single reactor could produce at the same production rate as two reactors in series with no increase in fouling rate.

The reactor was a boiling reactor where the heat of reaction was removed by the vaporization of the reactant, which was then condensed and returned to the reactor. The reaction product consisted of small, insoluble, solid particles. Because of the size of the particles, some were always entrained with the vapors leaving the reactor. The vertical condenser used to condense the reactant was provided with trays to allow the entrained solids to be washed back into the reactor. A sketch of this system is shown in Figure 12-1.

A plant test was deemed necessary since, at the higher vapor rates associated with the higher reactor capacity, more entrainment might be encountered. The increased entrainment might lead to fouling of the exchanger, the vapor line, or the liquid return line. In addition, the liquid handling capacity of the return line under possible fouling conditions was uncertain. It was necessary to develop techniques for monitoring the possible fouling of the exchanger, as well as the vapor line and liquid return line.

Three techniques were developed to monitor operations and fouling during the test of the single reactor. They were as follows:

1. The heat transfer coefficient on the overhead condenser was monitored hourly by the process control computer.
2. A local level gauge on the bottom of the exchanger was monitored every 2 hours by visual observations. An increase in the liquid level in the

bottom of the exchanger would indicate that the condensate was not flowing back into the reactor freely. That would mean that the liquid return line was likely partially plugged.

3. The vapor and liquid lines to and from the exchanger were "rung" every 2 hours. This "ringing" was a process of lightly tapping on the line with a hammer. If fouling was beginning to occur in the line, the pitch of the ring would have changed from a sharp, clear ring to a dull thud. While this qualitative test was not described in text books, it had been proven by operating personnel in earlier experiences.

The manual monitoring (local level gauge and "ringing the lines") required the operating personnel to climb six flights of stairs in order to reach the desired physical area of the equipment.

As indicated earlier, there are five aspects essential to a successful plant test. For this plant test, these items were covered as follows:

1. *Pretest Instrument and Laboratory Procedure Evaluation*: This was relatively simple for this test. The calibration of instrumentation that would allow personnel to determine the exchanger heat transfer coefficient was part of a routine task order. However, as indicated below, it was necessary to re-range several instruments to allow operation of the reactor at higher capacity. The manual procedures for checking the level and condition of the vapor and liquid lines were well established by previous experiences with solids entrainment.

2. *Statement of Anticipated Results*: Prior to the plant test, the heat transfer coefficient on the condenser was determined from the existing temperatures and heat duty. Calculations were made to determine what the anticipated inlet and outlet water temperatures would be at the anticipated heat duty. These calculations assumed that the heat transfer coefficient would be slightly lower than that determined at the existing duty. This was done to allow for a higher thickness of condensate flowing down the tubes. The actual change in condensate thickness would be very difficult to calculate with any accuracy. Thus the reduction in overall heat transfer coefficient caused by the thicker condensate layer was assumed to be 10%. Calculations were also made to confirm that the vapor and liquid lines had adequate capacity in their clean conditions. Operations personnel had an intuitive feel for what a clean line sounded like when it was rung. Thus an overall statement of anticipated results would indicate the following evaluation of the reactor overhead circuit was expected:
   - Based on the anticipated heat transfer coefficient, the anticipated values of the water temperatures and flow rates around the condenser were estimated. These were included in the anticipated results.
   - Based on calculations, there was anticipated to be no level in the bottom of the exchanger. In this case, the calculations indicated that

the liquid and vapor lines in a clean condition had sufficient capacity, so that there should be no level buildup in the exchanger.

- The overall statement included a comment that the vapor and liquid return lines should ring with a clear sound.

As with any process equipment or operating condition changes, safety must be considered. At the higher reaction rates, the safety valves would be required to release at a higher rate in an emergency. The required and actual capacities of the valves were reviewed and it was found that the capacity of the valves was greater than the maximum anticipated release rate. After other potential safety problems had been evaluated, a statement was provided indicating that no safety problems were anticipated.

Other areas of the process were also considered to evaluate the effect of operating at full rates with only a single reactor. An evaluation of the catalyst feed system to ensure that adequate catalyst supply would be available was conducted. The catalyst efficiency was anticipated to decrease as reactor volume was removed from service. All of the instrumentation required for the reactor to be operated at the higher rate was evaluated and meter ranges were increased where necessary.

3. *Potential Problem Analysis*: In the case of this plant test, there were potential problems associated with the overhead condenser as well as at other locations in the plant. Trigger points were developed for the heat transfer coefficient on the overhead condenser. This variable was routinely calculated by the process control computer and a trigger point was set to ensure that the reactor temperature did not run away. A reactor temperature runaway could occur if the heat transfer coefficient on the condenser is so low that a slight increase in reaction rate would cause the heat generated to increase at a faster rate than heat removal capabilities were able to handle. This was explained in detail in Chapters 4 and 10. The trigger points for the liquid level in the condenser and for the ringing of the lines were more qualitative. Any accumulation of liquid in the condenser or line ringing that was not clear were set as trigger points.

When operators focus on certain areas of the plant for a plant test, other areas are often overlooked. As indicated above, the catalyst feed rate was anticipated to increase due to the reduced residence time associated with using a single reactor. Since the reactor was a boiling reactor, some of the reactor volume below the liquid level was bubbling vapor. As such it would not be an effective reaction volume. Correlations were used to estimate the effect of the higher "boil up" rate associated with the increase in production on the bubbling vapor volume of the reactor. As the boil up rate increased, the correlations predicted that the bubbling vapor volume below the liquid level would increase and the effective reaction volume would decrease. The potential problem that these correlations were inaccurate was considered. If the reduction in reactor volume associated with the bubbling vapor was greater than expected, the catalyst efficiency would be lower than anticipated. Thus a trigger point was included for catalyst efficiency. Trigger points were determined for other parts of the process where potential problems were anticipated.

The key idea to consider is that having the well-thought-out potential problem analysis encourages reacting to situations in a predetermined way rather than in a reactionary mode in the midst of a problem.

4. *Explanation to Operating Personnel*: Most operating and mechanical personnel were happy to hear that their efforts to make a good product resulted in consideration of an increase in capacity. However, they had concerns that had to be addressed in order to obtain their cooperation for the plant test. Some of these comments and concerns were:

- "We don't have time to climb stairs and do the checks every two hours."
- "You guys will get all the credit while we do all the work."
- "Won't running the plant at a higher rate make more work for me?"
- "If we expand by adding more plants that will mean more jobs for my friends and family. This 'debottlenecking' expansion doesn't add any jobs."
- "Is it really safe to operate the reactor in this fashion?"
- "I don't see how we can run this reactor at higher rates when we are having problems with carryover at the lower rates."

Essentially, all of the questions that might come from the operating or mechanical personnel fall into two categories. There are questions that can be answered based on calculations that have already been done as part of the pretest efforts or potential problem analysis. The value of doing the calculations is that the operator can be told, "We have considered that and it is not a potential problem." The question of reactor safety is an example of this. It is always possible that a question will be raised which has not been considered. Serious consideration should be given to any question of this nature.

There will also be questions that cannot be answered based on calculations. Many of these will require an explanation that deals with company goals or the competitive situation. For example, the desire to expand by building more small plants could be answered by the fact that competitors are building larger plants and that being competitive with them provides job security. Job security for the employee is more important than creating jobs for friends and family.

5. *Formal Post-test Evaluation and Documentation*: While the test appeared to confirm the hypothesis, a formal test document was still prepared.

The formal test evaluation and documentation included the conclusion that the heat transfer coefficient of the condenser at the higher rates was comparable to that at the normal rates. The slightly higher condensate thickness did not cause any decrease in heat transfer coefficient. When the heat transfer coefficient was evaluated as a function of time, there was no indication that fouling was occurring during the test. In addition, there was no indication of fouling as gauged by the absence of any level in the condenser and the clear sound of the pipe ringing during the test.

The catalyst efficiency anticipated at the higher rates was compared to the actual catalyst efficiency. Since these values were very close, it was concluded that no unanticipated loss in reactor volume associated with vapor bubbling occurred during the test.

Perhaps one aspect that is often overlooked in documenting a plant test is the comments from mechanics or operating personnel. Comments of operating personnel were included. Their comments were that things ran very smoothly during the test.

Lessons Learned This test was successful from the standpoint that the results indicated conclusively that it would be possible to operate the reactors at significantly higher rates. Even if the conclusions had been that it would not be possible to operate the reactors at high rates, the test would still have been successful; the results would have been conclusive. As indicated earlier, a successful plant test is one where the test result is conclusive.

Several lessons were learned from this test. While the time to do the pretest calculations and potential problem analysis may seem to be inefficient, it provided a basis to convince management and operations personnel of the feasibility of the test. It also provided a basis for determining trigger points to allow a predetermined decision on aborting the test.

Using any short-term test to make decisions about fouling is difficult. The risk of concluding that fouling was not occurring was ameliorated by obtaining the maximum amount of data. In this case, three different techniques were used to monitor fouling.

There are always potential problems associated with any plant test. In the earlier example (12-1), the anticipation that the test was going to be boring was radically different than the anticipation for this test.

Even in tests that appear to confirm a hypothesis, documentation of the results is important. In this test, it was obvious that the results associated with fouling should be documented. In addition, the analysis of the expected catalyst efficiency at the higher rates and the higher boilup rate proved very valuable. Without this analysis and documentation, erroneous conclusions may have been reached regarding the reduced catalyst efficiency encountered as rates were increased.

While the discussion of pretest calculations indicated several technical procedures that a process operator might not have the training to perform, they are included here to illustrate what kind of calculations should be made prior to a plant test. Whether or not the process operator serving as a problem solver has the skills to make certain calculations, he should insist that the calculations be done prior to a plant test. An example of this in Example 12-2 is the calculation of the safety valve capacity as well as the safety valve release rates at the higher reactor production rates.

12.10 MORE COMPLICATED PLANT TESTS

It will frequently be necessary to consider conducting a plant test where the cost of a plant shutdown is so great that in-depth theoretical comparative studies must be done as part of the potential problem analysis. These studies will allow determining the risk of such an event by a comparison to existing operations. These comparative studies will not provide absolute values, but will allow determination of whether the plant test will increase or decrease the item of concern. An example of this might be a gas phase reaction process in which the presence of fines (small particles) could cause entrainment with a subsequent plant shutdown. Fines are known to be present with the current catalyst. However, the exact concentration is not known. The level of fines associated with the current catalyst results in minimal shutdowns. There are theoretical techniques available to predict the amount of fines produced by the catalyst, knowing the catalyst attributes (reactivity, particle size, and particle size distribution) and reaction conditions.

A plant test using a new catalyst with radically different attributes in the gas phase reactor was proposed. Concerns were expressed about the potential for this test to cause a plant shutdown. In this case, the absolute level of fines was not as important as answering the question of "Will the amount of fines increase or decrease?" In order to assess the risk of a plant shutdown if a plant test is run with the new catalyst, calculations were done to estimate the amount of fines produced by both the existing and new catalysts. These calculations were then compared and an assessment made regarding the possibility of increased fines. If this assessment indicated that there will be a comparable or lower level of fines produced, then there is a high probability that the plant test can be conducted without shutting down the plant. Conversely, if the calculations indicate that the fines level will increase, a plant shutdown is a definite concern. In this case, it might be better to redesign the catalyst rather than risk a plant shutdown caused by the test of a new catalyst.

12.11 DESIGN OF EXPERIMENTS (DOE)

During a laboratory investigation of a new process, new catalyst, or new product, it is often desirable to conduct a series of carefully planned experiments. This approach is referred to as a design of experiments (DOE). When one is working to solve a plant problem, the DOE approach is often unnecessary and is often counterproductive. There are three reasons for this:

- The range of variables in an operating plant is normally very broad. Thus data is available over almost any range that would be covered by a proposed DOE.
- DOEs are most applicable when there is no proposed hypothesis. Experiments must be conducted before one can even begin to develop a hypothesis. When one is solving a plant problem, the potential hypothesis is relatively easy to develop. But it must be proven by calculations and/ or a plant test.
- A true DOE will often cause a plant to operate at conditions that are outside of the stability limit, resulting in plant upsets or shutdowns.

12.12 KEY PLANT TEST CONSIDERATIONS

The key idea to remember from this chapter is that if a plant test is used to confirm the hypothesis, it must be well planned and aim to thoroughly prove or disprove the hypothesis. Successful plant tests are those that either confirm or reject the proposed hypothesis. Documentation of the plant test results, regardless of the conclusion, is also imperative.

13

UTILIZATION OF MANUAL COMPUTATION TECHNIQUES

13.1 INTRODUCTION

The purpose of this chapter is threefold:

- It provides an overview of the algorithm used by computers to do calculations such as flash and fractionation calculations.
- It includes techniques for doing these calculations by hand, as well as some shortcut approaches.
- It includes comments on the potential problems of computer solutions.

The modern-day widespread availability of desktop and laptop computers, along with process engineering programs, allows the chemical engineer or others to quickly simulate fractionation or flash calculations. In addition, the widespread availability of component data bases and equations of state/ equilibrium algorithms enhance the ability of the computers to do such calculations. Because of the expansion of the technology, it may seem that manual calculations are now obsolete. However, there is still a place for manual calculations. While the computer simulation programs are very precise, they also depend on accurate input data and good convergence routines. In addition, they are generally best suited for design rather than problem solving.

The use of any computer program also creates the risk of "user ignorance." That is, the user of the program simply will provide input data without

Problem Solving for Process Operators and Specialists, First Edition. Joseph M. Bonem.
© 2011 John Wiley & Sons, Inc. Published 2011 by John Wiley & Sons Inc.

understanding the algorithms used to perform the simulation. When using process engineering simulations, the program user must have a good understanding of the computer program to be used. This understanding has to be more than an understanding of how to fill in the blanks provided in the program. The user must understand how the algorithm works. Understanding of how to do the manual calculations will help provide the user's ability to comprehend the simulation programs.

The manual calculations discussed in this chapter include both of the following:

- Calculation techniques which were once well known and taught in the academic world, but have been forgotten due to lack of use.
- New techniques with which the problem solver may not be familiar.

The problems at the end of the chapter are given to illustrate these calculation techniques and are not in the format used in this book for example problems illustrating the five-step problem-solving approach.

Manual calculations are often of value as a "ball park" check on the computer simulation. Using the techniques discussed in this chapter will provide the means to perform checks on computer simulations. While manual calculations can never be as accurate as computer simulations, they provide a potential method to confirm that computer simulations are reasonably accurate. While the computer will not make calculation mistakes, it is subject to three sources of error as follows:

1. Programming errors. The simulation programmer may have made either a blatant or subtle error in the simulation. The blatant errors are almost always discovered and corrected prior to widespread use of the program. The subtle errors are often not detected until after months or years of use. An example of a subtle error is the simulation that uses numerical integration with a limited number of numerical increments. The simulation may work for years and give exact answers. However, it then fails to give the correct answer for a very specialized case that requires a greater number of increments than are used in the simulation. While this subtle error may be easily modified, the design and construction of the facilities based on the erroneous simulation may have already been completed.
2. The user of the simulation may have made a mistake in the input data. While this can also occur when doing manual calculations, the act of doing manual calculations will often allow discovery of simple input errors.
3. In his hurry to complete a computer simulation, the problem solver may compromise the validity of the solution by selecting inappropriate vapor-liquid equilibrium techniques, simulating a component not available in a data base with an inappropriate substitute, or accepting a less-than-acceptable convergence.

Unfortunately, once a computer simulation has been finished and presented, it has a great deal of credibility whether it is correct or not.

Manual calculations for binary systems can often be done quicker than computer simulations' calculations. This is especially true for the problem solver who often finds himself in a process plant control center without full computer simulation capabilities. This situation might also occur if the problem solver is in a country other than his own where he does not have access to a computer either due to security or language limitations. In these situations, and others where the process problem solver has available to him vapor pressure curves for the components under consideration, very good simulations of binary systems can be made by manual calculations. This area is discussed below in Sections 13.2 through 13.7.

Manual calculations can often provide good approximations of complicated equilibrium systems such those involving a polymer and a solvent. Many times the equilibrium between a polymer and a solvent is simulated by assuming that the polymer acts as the heaviest hydrocarbon available in the component data base. This assumption may be valid for high concentrations of solvent in the polymer. However, at low concentrations (almost always important when dealing with a solvent dissolved in a polymer) this is rarely true. This concept is discussed in section 13.8.

13.2   BINARY SYSTEMS

A binary system is one that consists of only two components. As indicated earlier, manual calculations for a binary system can be very accurate if vapor pressure curves or equilibrium constants are available for the components under consideration. Some areas where manual computation techniques can be done quickly and are of great value in binary systems are as follows:

- Estimating quickly the vapor and liquid compositions of a binary mixture.
- Estimating quickly the equilibrium of water that is dissolved in a hydrocarbon.
- Estimating quickly the vapor phase concentration when two liquid phases are present. The two liquid phases could be water and a hydrocarbon.
- Estimating quickly the condensation temperature of a gaseous mixture that contains both water and hydrocarbons.

The following paragraphs present examples of each area.

13.3   ESTIMATING THE VAPOR-LIQUID EQUILIBRIUM OF A TWO-PHASE BINARY MIXTURE

The phase rule given earlier and reproduced here can be used to understand how the composition of each phase of a binary system can be determined.

$$F = C - P + 2$$

$$(11-1)$$

where

F = degrees of freedom (temperature, pressure and composition)
C = number of components in the mixture
P = number of phases present

Thus for a binary ($C = 2$) vapor-liquid mixture ($P = 2$), fixing the temperature and pressure reduces the degrees of freedom to 0. Thus both liquid and vapor phase compositions are fixed.

For example, if a two-phase mixture of propane and isopentane exists at 150°F and 18 atmospheres, the composition of each phase can be calculated as follows:

1. Determine the equilibrium constant for each component. These can be determined either from charts of equilibrium constants as a function of temperature and pressure, or by using the relationship shown below:

$$K = VP/\pi$$

$$(13-1)$$

where

K = equilibrium concentration of the desired component
VP = vapor pressure of the component at 150°F
π = total pressure

2. Assign the following values:

X = composition of C_3 in the liquid phase, mol fraction
$1 - X$ = composition of I-C_5 in the liquid phase, mol fraction
Y = composition of C_3 in the vapor phase, mol fraction
$1 - Y$ = composition of I-C_5 in the vapor phase, mol fraction

3. Solve for $X, 1 - X, Y$, and $1 - Y$ using the following equations. The results are shown in Table 13-1:

$$Y_i = K \times X_i$$

$$(13-2)$$

$$\xi\, Y_i = 1.0$$

$$(13-3)$$

Component	Liquid, mf	K	Vapor, mf
C_3	X	1.28	$1.28 \times X$
I-C_5	$1 - X$	0.171	$0.171 \times (1 - X)$
Total	1.0		1.0

Table 13-1 Results of calculations

Component	Liquid, mf	Vapor, mf
C_3	0.748	0.957
$I\text{-}C_5$	0.252	0.043

thus

$$1.28 \times X + 0.171 \times (1 - X) = 1.0 \qquad (13\text{-}4)$$

Solving for X gives

$$1.28X - 0.171X = 1 - 0.171$$
$$X = 0.748 \qquad (13\text{-}5)$$

13.4 ESTIMATING THE EQUILIBRIUM CONSTANT OF WATER THAT IS DISSOLVED IN A HYDROCARBON

Estimating the volatility of water relative to the hydrocarbon in which it is dissolved is often required in process engineering calculations. The volatility is the ratio of the equilibrium constant of water to the equilibrium constant of the hydrocarbon. While it may seem apparent that water with a boiling point of 212°F is less volatile than lower molecular weight hydrocarbons such as propane and butane, this is not true if the water is dissolved in the hydrocarbon. Water is a unique molecule. Based on the atomic structure and molecular weight, it should be a gas under atmospheric conditions. The reason that water (molecular weight = 18) exists as a liquid rather than a gas is due to hydrogen bonding between the water molecules. The hydrogen atoms present in water (H_2O) associate with each other. This hydrogen bonding causes water to behave like a material of a much higher molecular weight and boil at a higher temperature. When water and hydrocarbons exist together, the water can exist either dissolved in the hydrocarbon or as a separate phase. The volatility of water associated with a hydrocarbon depends on whether the water is dissolved in the hydrocarbon or exists as a separate phase.

The volatility of water that is completely dissolved in hydrocarbon is greatly enhanced due to the breakdown of the hydrogen bonding. This breakdown results in water being more volatile than even low molecular weight hydrocarbons (e.g., propane). The enhanced volatility of water created by the breakdown of hydrogen bonding allows water to be removed from hydrocarbons, in some processes, by taking water as an overhead product. Such a drying by fractionation process is shown in Figure 13-1.

Often it is desirable to estimate the equilibrium constant and, hence, volatility of water dissolved in a hydrocarbon to allow design of a distillation column such as that shown in Figure 13-1. Some simulation programs can correctly handle water that exists as a second phase or dissolved in the hydrocarbon.

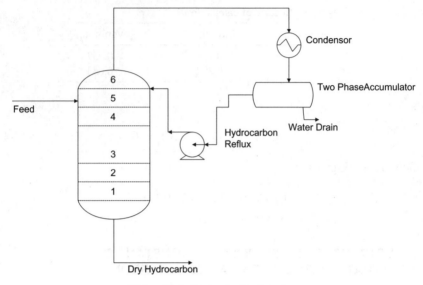

Figure 13-1 Drying by fractionation.

The volatility of water can also be estimated by manual calculations. This can be done knowing that the activity of a component in all phases is equal and assuming that the hydrocarbon is essentially saturated with water. Activity is a property of any chemical component. It is a function of temperature and composition.

This can be expressed mathematically, as follows:

$$A_W = A_V \tag{13-6}$$

$$VP = PP = Y \times \pi \tag{13-7}$$

$$Y = K \times X \tag{13-8}$$

by substitution

$$K = VP/(X \times \pi) \tag{13-9}$$

where

A_W = activity of water in the hydrocarbon phase if it is dissolved in the hydrocarbon

A_V = activity of water in the vapor phase

K = equilibrium constant of water dissolved in any hydrocarbon

VP = vapor pressure of pure water at the chosen temperature

· π = total pressure which will be approximately the vapor pressure of the hydrocarbon

PP = partial pressure of water vapor

Y = concentration of water in the vapor phase, mol fraction

X = saturation concentration of water in hydrocarbon at the chosen temperature, mol fraction

Equation (13-9) indicates that the equilibrium constant for water in any hydrocarbon at saturation can be estimated knowing only the saturation concentration and the vapor pressure of water and the hydrocarbon. This approach is based on the following assumptions:

- Since the water concentration in the hydrocarbon phase is assumed to be at saturation (that is, a single additional drop will cause formation of a separate water phase), the vapor pressure of water can assumed to be that of pure water. The vapor pressure should be very close to the activity. These assumptions allow the translation of the left hand side (LHS) of equation 13-6 to the LHS of equation 13-7.
- As equation 13-7 shows, the activity is assumed to be equal to the partial pressure or the vapor phase concentration multiplied by the total pressure.

While the expert in thermodynamics may consider the approach described above less than desirable, it is more than acceptable for plant problem solving.

As indicated above, this approach is applicable for estimating the equilibrium constant of water dissolved in a hydrocarbon when the hydrocarbon is saturated with water. If the concentration is below the saturation level, the equilibrium constant for water will likely be even higher. This is expected because as the concentration of a component in a nonideal system decreases, the deviation from ideal increases.

13.5   ESTIMATING THE VAPOR PHASE CONCENTRATION WHEN TWO LIQUID PHASES ARE PRESENT

Two immiscible liquid phases are often present within hydrocarbon and water mixtures. It is often desirable to estimate the vapor phase composition in equilibrium with the two liquid phases. An example of this might be a knockout drum that contains both water and a hydrocarbon. If the vapor phase from the drum is to be dried in an adsorption bed, it will be necessary to know the water content of the vapor leaving the knockout drum. Such a process is shown in Figure 13-2.

The vapor phase composition can be estimated by the simple rule that each liquid phase exerts its own vapor pressure. That is, since the liquid phases are

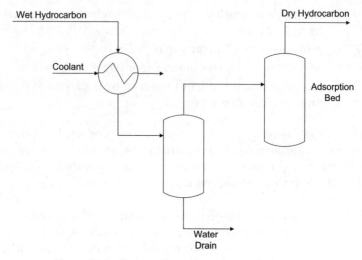

Figure 13-2 Drying with condensation and adsorption.

immiscible, they act independently and the vapor pressure of either depends only on the temperature.

Thus

$$\Pi = VP_W + VP_{HC} \tag{13-10}$$

here

Π = total pressure of the two liquid phases or the total pressure of the system

VP_W = vapor pressure of the water phase

VP_{HC} = vapor pressure of the hydrocarbon phase

For any given system, if the total pressure (π) and vapor pressures of the hydrocarbon and water are known, then the vapor phase composition in mol fraction can be calculated as follows:

$$Y_W = VP_W/\pi \tag{13-11}$$

$$Y_{HC} = VP_{HC}/\pi \tag{13-12}$$

here

Y_W = mol fraction water in the vapor phase

Y_{HC} = mol fraction hydrocarbon in the vapor phase

13.6 CONDENSATION INTO TWO LIQUID PHASES

A converse of Section 13.5 is the case where a vapor phase containing both a hydrocarbon and water is condensed. In this situation, the temperature of the vapor is reduced until condensation of the vapor begins. Both the initial condensation temperature and which phase condenses first (water or hydrocarbon) depend on the initial composition of the vapor phase. Determination of which phase will condense first, the initial condensation temperature, and the subsequent temperature at which formation of a second liquid phase occurs can be determined by the single process engineering concept as follows:

> *The partial pressure of a component in the vapor phase can never exceed the vapor pressure of the component at the given temperature.* If the partial pressure exceeds the vapor pressure, condensation must occur.

Thus one of the components (water or hydrocarbon) will begin to condense at a temperature of T_1 when the partial pressure exceeds the vapor pressure. This can be expressed mathematically as follows:

$$\Pi \times Y_i \leq VP_i \qquad (13\text{-}13)$$

here

Y_i = vapor phase composition of the first component to condense

VP_i = vapor pressure of the first component to condense

The initial component will continue to condense by itself until the temperature of the vapor is reduced to the point (T_2) where the sum of the vapor pressures is equal to the system pressure. At this point, equation (13-10) is satisfied and condensation of the second component will begin. The temperature (T_2) will stay constant until the entire second component is condensed. At this point, the there will be no vapor left to condense. Additional heat removal will serve to cool the liquid. The vapor phase composition of component "i" (the first component to condense) between T_1 and T_2 can be estimated knowing the temperature as follows:

$$Y_i = VP_i / \pi \qquad (13\text{-}14)$$

An example problem is provided later to enhance the understanding of this concept.

It should be noted that the concept described above, that the partial pressure of a component in the vapor phase can never exceed the vapor pressure of the component at the given temperature, can also be used to evaluate the validity of plant laboratory or instrument data. For example, vapor phase analytical data must satisfy equation (13-13). The term $\Pi \times Y_i$ cannot exceed VP_i. If it does exceed VP_i, it is likely that composition is wrong.

13.7 SHORTCUT FRACTIONATION CALCULATIONS

Computer fractionation simulation programs can be used to both design new distillation towers and to solve problems with existing towers. Most of the simulation programs are intended for design calculations. They can be used for problem solving, but to operate in this mode often requires iterative procedures. For example, the design of a distillation column requires an input of feed and product compositions, rates, tray efficiency, and a criterion for setting the reflux ratio. The reflux ratio is generally set at a multiple of the minimum reflux ratio which is calculated by the computer simulation program. The simulation program can then size the tower. If problem solving is being done on the same tower, the reflux ratio and the number of actual trays are fixed. The tray efficiency must be adjusted to obtain a match between calculated and actual product compositions.

Even though the computer simulation programs are used for the fractionation calculations, there are some advantages to knowing shortcut fractionation techniques. Knowing these techniques will aid in understanding the results of the simulation as well as being able to check the simulation results. Being familiar with shortcut calculation techniques will also be of value on occasions when a quick answer is required to an immediate problem. As indicated earlier, the computer will not make computational errors, however, it will only do what it is told to do. There are three types of errors that might occur in any simulation. These were described in Section 13.1.

Shortcut fractionation calculations can be used to confirm that computer simulations are giving "directionally correct" answers. In addition, these shortcut calculations can often be used to analyze poor performance in an operating tower in a more expeditious fashion than is possible with a full computer simulation program. These shortcut calculations can be set up on a spreadsheet with a minimal amount of effort, as opposed to computer simulations which are quick to run, but take time to set up and develop input data in the correct format.

One of the most valuable parts of a computer simulation program is its capability to determine the minimum reflux ratio for a multicomponent system. Since, in a plant operating column, the reflux ratio is already known, determining the minimum reflux ratio may not be required. Prior to launching a rigorous simulation of an existing fractionation column, a shortcut calculation procedure should be considered. The possible reasons for considering a manual shortcut calculation approach are as follows:

- The shortcut calculation approach is much less complicated.
- The reflux ratio in an existing column is already known and the minimum reflux ratio generally does not need to be determined.
- A plant tower is often a binary fractionation (it has only two components) or can be simulated as one.
- A plant problem often involves only part of the tower (rectification or stripping sections); a complete simulation is not required.

- For ideal systems (such as systems involving only hydrocarbon), a heat balance around each tray is not required since "equal molal overflow" is a good assumption. The assumption of equal molal overflow means that the internal liquid and vapor rates expressed in mols/hour will be equal throughout the tower if there is no addition or withdrawal of material or heat.
- The simplified vapor-liquid equilibrium constants that are used for the shortcut calculations are often much easier to develop than the more complex ones often required for the computer simulations.

Even if a rigorous simulation program is used to predict plant performance or to design a new tower, it will be of value to perform a shortcut calculation to confirm the validity of the simulation. As an example of a shortcut calculation, the rectification section of a tower is shown in Table 13-2. The calculation bases are given in steps 1 and 2 (equations (13-15) and (13-16)). This shortcut procedure assumes that the vapor-liquid equilibrium is constant in the tower and that equal molal overflow is a good assumption. A spreadsheet was developed and the results are shown in Table 13-3.

Note that if the calculated composition does not equal the assumed composition, a new assumption must be made and the calculations repeated.

The calculations can be performed in a stepwise fashion using the two equations shown below. These are:

1. An equilibrium relationship between vapor and liquid on any tray

$$Y_N = \alpha \times X_N / (1 + (\alpha - 1) \times X_N) \qquad (13\text{-}15)$$

2. A material balance relationship between vapor and liquid and overhead distillate product,

---

Table 13-2   Shortcut calculations to simulate rectification section

---

Given:

- Number of theoretical rectification trays = 15. These are the number of theoretical trays from the feed tray or lowest tray in the section to the overhead product.
- Relative volatility (light key to heavy key) = 1.2. This value is simply the ratio of the equilibrium constant of the more volatile component (light key) to the less volatile component (heavy key).
- Internal liquid to vapor ratio, mol/mol = 0.9. This is the number of mols of liquid flowing down the tower divided by the number of mols of vapor flowing up the tower.
- Liquid concentration of light key on lowest tray, mol % = 70.
- Assumed concentration of light key in distillate, mol % = 93.

Table 13-3 Results of shortcut calculation

Tray Number	Compositions of Light Key in Mol Fraction	
	Liquid on Tray	Vapor leaving Tray
1 (starting tray)	0.7	0.737
2	0.715	0.751
3	0.731	0.765
4	0.747	0.78
5	0.763	0.795
6	0.78	0.809
7	0.796	0.824
8	0.812	0.839
9	0.828	0.853
10	0.844	0.867
11	0.86	0.88
12	0.875	0.893
13	0.889	0.906
14	0.903	0.918
15 (top tray)	0.917	0.93

$$X_{N+1} = Y_N/R - X_D \times ((1/R) - 1) \qquad (13\text{-}16)$$

where

X_D = distillate composition of light key. For simple fractionation towers, it is the same composition as the vapor leaving the tower since all of this is condensed and becomes the distillate product and tower reflux.

Y_N = vapor phase composition of light key on tray N, mol fraction.

X_N = liquid phase composition of light key on tray N, mol fraction.

X_{N+1} = liquid phase composition of light key on tray above tray N, mol fraction.

α = relative volatility, light key to heavy key.

R = internal liquid to vapor mol flow rate ratio.

Light key refers to the component that is the most volatile; heavy key refers to the component that is the least volatile.

The actual procedure for doing these calculations, whether by hand or in the computer, is as follows:

1. Using equation (13-15) and the given relative volatility ($\alpha = 1.2$), calculate the vapor phase composition leaving the lowest tray based on the given composition of 70 mol % for X_N.

2. Using Y_N calculated in Step 1, the internal liquid to vapor ratio (R), the light key composition in the distillate (X_D), and equation (13-16), calculate the liquid phase composition leaving the tray immediately above (X_{N+1}).

3. Repeat this calculation approach for each of the theoretical trays. If the vapor leaving the 15th tray does not equal the assumed distillate composition (93 mol % in this case), it will be necessary to assume a new distillate composition and redo the calculations.

Table 13-3 summarizes the results of the spread sheet calculations.

Since the vapor leaving tray 15 is condensed and is the distillate, the calculated composition equals the assumed composition.

These calculations can be simplified even further for a binary tower at constant temperature where *small quantities of a heavy key (lower volatility material) are being fractionated to low levels in the overhead product.* In this case, it can be shown that:

$$X_{n+1} = (X_n \times V \times k)/L \qquad (13\text{-}17)$$

where

X_{n+1} and X_n = tray compositions of the heavy key in mol fraction, where tray $n+1$ is above tray n

V = internal vapor rate in mol/hr

L = internal liquid rate in mol/hr

k = volatility of the heavy key relative to the light key. Note that this will always be less than 1.

Equation (13-17) can be transformed into the equation below, which can be used to make a quick estimate of the low volatility material in the distillate product.

$$X_D = X_T \times (V \times k/L)^N \qquad (13\text{-}18)$$

where

X_D = composition of the heavy key in mol fraction in the overhead distillate product

X_T = composition of the heavy key in mol fraction at the start of the rectification section

N = number of theoretical trays in the rectification section

13.8   POLYMER-SOLVENT EQUILIBRIUM

The problem solver will often not have available to him simulations that allow predictions of polymer-solvent equilibrium. Faced with this dilemma, he often will select the heaviest component in the computer data base to simulate a polymer. This may be adequate for solvent-polymer systems that contain solvent concentrations greater than 10 to 20 volume percent. However, below this level, the polymer begins to reduce the volatility of the solvent, and the vapor phase composition of the solvent will be less than anticipated when the polymer is simulated using the heaviest component in the data base. The converse will also be true. That is, at solvent concentrations below 10 to 20 volume percent, the amount of solvent in the polymer will be greater than that calculated based on the vapor phase composition of the solvent and simulating polymer as the heaviest component in the data base.

There are techniques for dealing with this nonideality. Perhaps the best known relationship dealing with the equilibrium between a polymer and solvent is the Flory-Huggins relationship. The use and limitations of this relationship are described in *Seymour/Carraher's Polymer Chemistry*.

The Flory-Huggins relationship predicts the equilibrium of a solvent in the noncrystalline polymer phase. Polymers exist in either the crystalline or noncrystalline (also referred to as amorphous) phase. The crystalline phase can be visualized as a piece of metal. Since all the molecules are aligned in the available, specific places, there is no location available for solvent molecules. Hence the solvent molecules cannot be present in the crystalline region. This is not true for the noncrystalline or amorphous phase. Thus any solvent present in a polymer is assumed to be present in the amorphous phase of the polymer. Typical polyolefins, for example, have a crystalline phase that is equal to 70–90% of the polymer.

Equation (13-19) below shows the original Flory-Huggins relationship. The modification developed by the author is shown in equation (13-20).

The original Flory-Huggins relationship:

$$\ln(PP/VP) = \ln V_1 + (1 - M_1/M_2) \times V_2 + U \times V_2^2 \qquad (13\text{-}19)$$

where

PP = solvent partial pressure; this is equal to the vapor phase composition multiplied by the total system pressure

VP = vapor pressure of solvent

V_1 = volume fraction of solvent in polymer

V_2 = volume fraction of polymer

M_1 = molar volume of solvent, cc/g-mol

M_2 = molar volume of polymer, cc/g-mol

U = interaction parameter between the solvent and the polymer

Manipulation and assumptions ($V_2 = 1$ and $M_1/M_2 = 0$) by the author of this book yield

$$\ln(PP/VP) = \ln(D_2 \times X_1/(D_1 \times 1,000,000)) + 1 + U \qquad (13\text{-}20)$$

where

D_1 = density of the solvent
D_2 = density of the polymer
X_1 = weight of the solvent in the polymer, ppm

The interaction parameter can be evaluated knowing the solubility parameters.

$$U = M_1 \times (S_1 - S_2)^2/(R \times T) + Z \qquad (13\text{-}21)$$

where

S_1 = solvent solubility parameter, $(cal/cm^3)^{0.5}$
S_2 = polymer solubility parameter, $(cal/cm^3)^{0.5}$
Z = lattice constant. In theory = 0.35 ± 0.1

The values of solubility parameters are given in various handbooks. It should be emphasized that the assumed concentrations of solvent in the polymer are the concentrations of solvent which exists in the noncrystalline or amorphous region. That is, the standard assumption is that there is no solvent in the crystalline region.

While this book cannot delve into all aspects of polymer-solvent equilibrium, the approach described above is meant to show that there are techniques available to estimate the equilibrium as opposed to simulating a polymer as the heaviest component in the data base. While there are computer simulations that allow prediction of a polymer-solvent equilibrium, the above analysis will be of value in understanding the basis for these simulations or in doing the simulation by manual calculation.

13.9 EXAMPLE PROBLEMS

The following example problems illustrate the application of these concepts. These are shown to illustrate the calculation approach. As such, they are not formatted in the style of the standard problem-solving technique proposed in this book.

EXAMPLE PROBLEM 13-1

A two-phase liquid mixture consisting of hexane and water is present in a process vessel at a pressure of 5 psig. The mixture enters the vessel at the

Table 13-4   Vapor pressure data

Component	Temperature, °F					
	100	120	140	160	180	200
	Vapor Pressure, psia					
Water	0.95	1.69	2.89	4.74	7.51	11.53
Hexane	5	7.79	11.02	15.4	22	28.7

Table 13-5   Boiling point estimate

	Temperature, °F			
	100	120	140	160
VP of water, psia	0.95	1.69	2.89	4.74
VP of hexane, psia	5	7.79	11.02	15.4
Total (π),[a] psia	5.95	9.48	13.91	20.14
Mols water/mol hexane	0.19	0.22	0.26	0.308

[a] By interpolation, π will equal 19.7 (5 psig + 14.7) when the temperature of the mixture is equal to 159°F. This is, by definition, the boiling point.

boiling point. What is the boiling point of this mixture? The vapor from this drum must be dried to remove water. How could the amount of water present in the vapor phase be reduced below 0.26 mols per mol of hexane? The vapor pressure data in psia is given in Table 13-4.

Since there are two liquid phases present and the mixture is at the boiling point, equation (13-10) shown below provides the basis for the solution.

$$\Pi = VP_W + VP_{HC} \qquad (13\text{-}10)$$

where

Π = total pressure of the two liquid phases

VP_W = vapor pressure of the water phase

VP_{HC} = vapor pressure of the hexane phase

Using the vapor pressures given earlier, the boiling point of the mixture can be determined as shown in Table 13-5.

The vapor phase compositions expressed as mols of water/mol of hexane in Table 13-5 were developed remembering that:

- The partial pressure of a component is equal to the vapor pressure multiplied by the mol fraction of the component in the liquid. Since the hexane and water exist as separate phases, each has a mol fraction in the liquid of 1. Thus, in this case, the partial pressure is equal to the vapor pressure of each component.

- The ratio of the mols of each component is then simply the vapor pressure ratio.

From an examination of Table 13-5, if the vessel were operated at a slight vacuum (<13.9 psia), the amount of water in the vapor would be below 0.26 mols of water/mol of hexane.

EXAMPLE PROBLEM 13-2

A vapor phase mixture at 200°F and 20 psia, containing 50 mol percent water and 50 mol percent hexane, flows into a cooler. Determine the following using the vapor pressures given in the previous problem:

- The temperature of the initial condensation and which component condenses first.
- The temperature at which the remaining component begins to condense.

The solution of this problem depends on utilizing the concepts discussed earlier. That is, *the partial pressure of a component can never exceed the vapor pressure of the component at the given temperature.*
Mathematically, this was expressed as:

$$\Pi \times Y_i \leq VP_i \tag{13-13}$$

The partial pressure of both hexane and water is 10 psia, since the vapor phase contains 50 mol percent of each component. As noted in equation (13-13), the vapor pressure must always be equal to or greater than the partial pressure. Since the mixture is being condensed, the vapor pressure and partial pressures will be equal. Interpolating the vapor pressures given earlier, it can be shown that a vapor pressure of 10 psia occurs at the temperatures shown in Table 13-6.

Since as the temperature is decreased, the condensation point of water occurs first, it will form the initial liquid phase and it will begin to condense at 192°F. It is important to recognize that the calculation shown above is valid only for finding the component which condenses first and its condensation point.

Hexane will not begin to condense until the total vapor pressure of the two liquid phases equals the total pressure (20 psia). This can be estimated as shown in Table 13-7.

Table 13-6   Estimation of condensation point

Component	Temperature, °F for $VP = 10$ psia
Water	192
Hexane	134

Table 13-7   Estimation of condensation point of second phase

Component	Temperature, °F		
	120	140	160
Water, psia	1.69	2.89	4.74
Hexane, psia	7.79	11.02	15.4
Total, psia	9.48	13.91	20.14

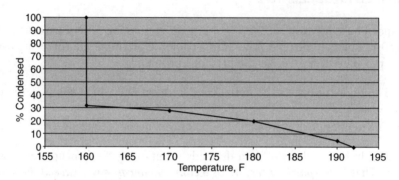

Figure 13-3   Percentage condensation vs. temperature.

Thus at a temperature ≈ 160°F, sufficient hexane will condense to form a second phase. At this point (160°F), the vapor phase composition can be estimated from the vapor pressures by dividing the water vapor pressure by the hexane vapor pressure, which gives a composition of 0.308 mol of water per mol of hexane.

If a heat exchanger to do this condensation job were being designed, it would be split into different zones. From 200°F to 192°F, there will be no condensation. This is a vapor cooling zone only. From 192°F to 160°F, only water will condense. There will be trace amounts of hexane that condense as necessary to saturate the water phase with hexane. The vapor phase composition in this range can be estimated based on equation (13-22) as follows:

$$Y_W = VP_W/\pi \tag{13-22}$$

where

Y_W = mol fraction of water in the vapor phase

VP_W = vapor pressure of water at the temperature of interest, psia

Π = total pressure, psia

When a temperature of 160°F is reached, hexane will begin to condense. The temperature will remain at 160°F until all hexane and water are condensed. A condensation curve for this system is shown in Figure 13-3.

EXAMPLE PROBLEM 13-3

Find the approximate relative volatility for water dissolved in propane at 70°F.
Given:

- Vapor pressures at 70°F:

Water	0.363 psia
Propane	118 psia

- Solubility of water in propane at 70°F = 0.016 lb/100 lb propane

Calculations:
   Converting the solubility to mol fraction gives 0.00039 mol water/mol propane

Since essentially all of the pressure on the system is due to propane

$$\Pi = 118\,\text{psia}$$

Substituting in (13-9) gives the following results:

$$K = VP/(X \times \pi) \tag{13-9}$$
$$K = 0.363/(0.00039 \times 118) \tag{13-23}$$
$$K = 7.88 \tag{13-24}$$

Since the equilibrium constant for propane is 1 (vapor pressure/total pressure), the relative volatility of water to propane is approximately 8. Thus water would be more volatile than propane.

NOMENCLATURE

A_V Activity of water in the vapor phase

A_W Activity of water in the hydrocarbon phase if it is dissolved in the hydrocarbon

C Number of components in the mixture

D_1 Density of the solvent

D_2 Density of the polymer

F Degrees of freedom (temperature, pressure, and composition)

K Equilibrium concentration of the desired component. In this chapter it is used in reference to a hydrocarbon or water dissolved in a hydrocarbon.

k Volatility of the heavy key relative to the light key. Note this will always be less than 1.

L The internal liquid rate in a fractionation tower, mol/hr

M_1 Molar volume of solvent, cc/g-mol

M_2 Molar volume of polymer, cc/g-mol

N Number of theoretical trays in the rectification section

P Number of phases present

PP Partial pressure of the component under study

R Internal liquid to vapor mol flow rate ratio in a fractionation tower

S_1 Solvent solubility parameter, $(cal/cm^3)^{0.5}$

S_2 Polymer solubility parameter, $(cal/cm^3)^{0.5}$

U Interaction parameter between the solvent and the polymer

V Internal vapor rate in a fractionation tower, mol/hr

V_1 Volume fraction of solvent in polymer

V_2 Volume fraction of polymer

VP Vapor pressure of the component under study. In this chapter, it could be the vapor pressure of the hydrocarbon phase (VP_{HC}), the vapor pressure of the water phase (VP_W), the vapor pressure of the first component to condense (VP_i) or the vapor pressure of a solvent dissolved in a hydrocarbon (VP).

X Concentration of a component in the liquid phase, mol fraction. As used in this chapter it could be either the saturation concentration of water in hydrocarbon at the chosen temperature or the hydrocarbon concentration in a mixture.

X_D Composition of the heavy key in mol fraction in the overhead distillate product

X_N Liquid phase composition of light key or heavy key on tray N, mol fraction

X_{N+1} Liquid phase composition of light key or heavy key on tray above tray N, mol fraction

X_T Composition of the heavy key in mol fraction at the start of the rectification section

X_1 Weight of the solvent in the polymer, ppm

Y Concentration of a component in the vapor phase, mol fraction. It could be water (Y_W), hydrocarbon (Y_{HC}), or the composition of the first component in a hydrocarbon-water mixture to condense (Y_i).

Y_N Vapor phase composition of light key or heavy key on tray N, mol fraction

Z Lattice constant. In theory $= 0.35 \pm 0.1$.

α Relative volatility, light key to heavy key

Π Total pressure

Note   Light key refers to the component that is the most volatile. Heavy key refers to the component that is the least volatile.

14

PUTTING IT ALL TOGETHER

14.1 INTRODUCTION

The previous chapters have focused on various phases of problem-solving procedures and activities as well as process engineering calculation techniques. The actual procedures and techniques applicable for any problem-solving activity have been discussed. In addition, Chapters 7–13 contain experience-based process engineering calculation techniques and guidelines. In each of these chapters, there are example problems that are directed toward the theme of the particular chapter. In real life, problems don't come packaged so neatly. For example, what is described as a problem associated with a prime mover may well be a problem associated with a reactor.

In order to see how these areas fit together, several real-life example problems are discussed in this chapter. The chapter shows how the five-step problem-solving procedure (Chapter 3) can be used, how working hypotheses can be formulated (Chapter 6), and how various process engineering calculations (Chapters 5 and 7–13) can be utilized to develop and confirm these hypotheses.

14.2 DON'T FORGET TO USE FUNDAMENTALS

Two of the most powerful tools that a process engineer (or a problem solver serving as a process engineer) can use in problem solving are heat and material

Problem Solving for Process Operators and Specialists, First Edition. Joseph M. Bonem.
© 2011 John Wiley & Sons, Inc. Published 2011 by John Wiley & Sons Inc.

balances. These concepts were described in Chapter 5. The material balance simply states that, with the exception of atomic power, mass cannot be created or destroyed. Thus the total flow in mass units into a process or a unit operation must be equal to the total flow out. A comparable truism applies to heat balances. In addition to the need for a process to be in balance, heat and material balances can be used to determine unknowns. For example, material balance principles can be used to determine the production rate of a desirable or undesirable component. If an undesirable byproduct is being produced in a reactor and removed in a purge stream, the rate of production is simply the removal rate. Heat balances can be used to determine the boilup rate from the steam rate. A problem solver might be told that a tower is flooding because the trays are plugged. When he performs a heat balance, he might conclude that the heat input to the tower is less than the heat being removed from the tower. If he makes a closer examination, he might find that the problem really is a steam meter that is indicating a flow rate much lower than the actual flow. This is causing the tower to be flooded, due to the excessive vapor being generated in the reboiler.

As will be noted in the example problems, knowing the flow rates of process or utility streams is a requirement for successful problem solving. While this may seem very basic, it is amazing how many problem solvers will start developing intricate hypotheses even though flow instruments indicate that the amount of material coming into the unit operation does not equal the amount of material leaving. Thus one of the key ideas for a problem solver to remember is to not forget the fundamentals of their discipline.

EXAMPLE PROBLEM 14-1

Do Fundamental Processes Developed in the United States Translate to Europe?

A chloride removal unit designed and operated in the United States was an integral part of a gas-drying process that used triethylene glycol (TEG) as the circulating drying solvent. Water was removed from the gas in the absorber using a stream of TEG that had a very low water content. The properties of TEG are such that water has a very low partial pressure when dissolved in the TEG. Thus it can be readily used to dry gases of all descriptions.

In the drying process, the TEG leaving the absorber and containing the water removed in the absorber flowed to a heated, two-stage regeneration system. In this regeneration system, the dissolved water was stripped from the gas and the dry TEG was recirculated to the absorber. The initial stage of the regeneration system was a heated flash step with about 5 min of residence time. The second stage was a vacuum distillation tower. The TEG from the vacuum distillation tower flowed back to the absorber.

The gas to be dried was a chlorine derivate. Some of the gas dissolved in the circulating TEG was converted to hydrogen chloride (HCl) when the solution was heated as part of the regeneration step. The reaction of the gas to form HCl was known to be first order with respect to the concentration of the chlorine derivative in the circulating TEG. However, the exact reaction rate constant and Arrhenius constant were not known. The Arrhenius constant is a value that describes how fast the reaction increases with increasing temperature. Essentially, all of this reaction occurred in the first-stage flash drum of the two-stage regeneration system. This stage was operated at 175°F and 5 psig. The level in the drum was controlled so that the residence time was limited to 5 min. The second stage of the regeneration system consisted of a small vacuum fractionating tower. HCl generation in this stage was very limited because of the short residence time and low pressure.

The HCl formed was neutralized using a soluble amine. However, the amine could not be added continuously since the HCl-amine complex would build up in the circulating TEG to an unacceptably high level. The HCl was removed from the HCl-amine complex by passing a small stream of dry TEG through a drum filled with ion exchange resin. Ion exchange resins are complex salts supported on a synthetic resin such as polystyrene. This particular ion exchange resin had the capability of removing the Cl⁻ ion and replacing it with an OH⁻ ion. Thus the HCl-amine bond was broken and the HCl was converted to water. The ion exchange resin itself was converted from an OH⁻ form to a Cl⁻ salt. As this process continued, the bed would eventually become saturated with Cl⁻ ions and become ineffective. To restore the bed, operators removed it from service and regenerated it using water and a sodium hydroxide solution. This regeneration was done whenever the concentration of chlorides in the outlet increased above 50 ppm. A simplified drawing of the process is shown in Figure 14-1.

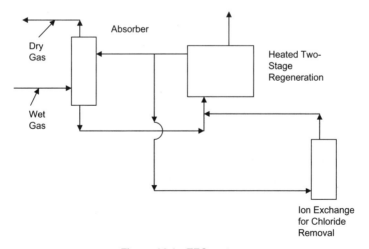

Figure 14-1   TEG system.

The process was designed and installed in a location in the United States. After initial startup problems, the operation of the process was very successful. The chlorides were controlled at a concentration of 500 ppm with a flow to the ion exchange bed of 500 lb/hr. During the initial operations, high corrosion rates where encountered when the chloride level increased to 1000 ppm.

An identical unit was installed in Europe. Flow rates, temperatures, and pressures were all the same. While the gas was successfully dried, the concentrations of chlorides in the circulating TEG rose to 1500 ppm in a month after startup. The ion exchange bed appeared to be operating well, since the outlet concentration from the bed was zero until the ion exchange resin became saturated with chlorides. However, the bed had to be regenerated more frequently than anticipated due to the heavy loading of chlorides. At the time, the plant in Europe was not using the problem-finding concepts discussed in Chapters 3 and 6. Because of the concern over corrosion, European management shut the unit down. While significant corrosion had not yet been observed, they were convinced, based on experience in the United States, that high corrosion rates would be observed soon. After some preliminary problem-solving attempts on their own, they requested help from the original designer in the United States.

When the problem solver from the United States arrived on the scene, he began a methodical problem-solving activity using the approach discussed in the previous chapters. This methodical problem-solving approach was even more important in this case since there were significant geopolitical factors involved. There was a great deal of animosity between the European affiliate and the U.S.-based technical staff. The problem solver began the five-step approach as described in Chapter 3.

Step 1: Verify that the problem actually occurred.

Since operation at high chloride concentrations had not yet resulted in observable corrosion, the first problem to be verified was that the actual concentration of chlorides was as high as 1500 ppm. The laboratory procedure was confirmed and several samples were analyzed confirming that, indeed, the chloride concentration was 1500 ppm. Since the corrosion rate had been so severe in the United States at a chloride level of 1000 ppm, it did not seem wise to continue operation and simply monitor the corrosion rate.

The problem solver's next step in verifying that the problem actually occurred was to compare the theoretical chloride buildup to the actual chloride buildup in Europe. Since the reaction rate constant was not known, the theoretical chloride buildup was determined based on operations in the United States. Based on material balance principles, the problem solver calculated the chloride production rates in the United States. This material balance principle is as follows:

$$RC = F \times X_F \qquad (14\text{-}1)$$

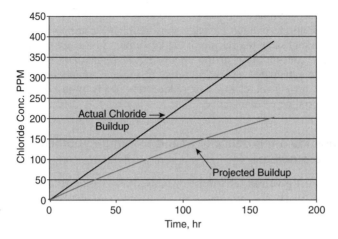

Figure 14-2   Actual and projected chloride buildup.

where

RC = rate of chloride production in the United States, lb/hr

F = flow rate to the ion exchange bed, lb/hr

X_F = concentration of chlorides in the flow to the ion exchange bed, weight fraction

Note that equation (14-1) assumes that the chloride concentration in the outlet from the ion exchange bed was zero. This was true essentially all of the time.

Knowing the rate of chloride production in the United States and assuming that it was the same in Europe, the problem solver could estimate the chloride concentration buildup rate in the European plant by knowing the TEG inventory. He then developed the relationships shown in Figure 14-2. In this figure, the projected chloride buildup based on kinetic relationships developed in the United States is shown along with the actual chloride buildup. This provides a means of comparing the theoretical and actual chloride concentration buildup rates. Obviously, there was a significant difference between the two rates.

Step 2: Write out an accurate statement of what problem you are trying to solve.

The problem solver wrote out the problem statement as follows:

The chloride concentration in the circulating TEG in the European plant has built up to 1500 ppm instead of the anticipated level of 500 ppm. This is a major concern because experience in the United States indicated that significant corrosion would begin at levels above 1000 ppm. This high level of chlorides is being experienced in Europe at the apparent identical process conditions used in the

United States, which resulted in a chloride level of only 500 ppm. An analysis of the rate of chloride buildup indicates that the higher chloride production rate has been present since the startup of the equipment in Europe. Determine why the chloride concentration in the circulating TEG has built up to 1500 ppm instead of an anticipated level of 500 ppm. In addition to understanding what is causing the high level, changes to reduce the steady state concentration to 500 ppm should be recommended.

Step 3: Develop a theoretically sound working hypothesis that explains the problem.

The question list given in Chapter 6 was used as a guide to develop potential working hypotheses. A summary of this analysis is shown below, in Table 14-1.

Table 14-1 Questions/comments for Problem 14-1

Question	Comment
Are all operating directives and procedures being followed?	All appeared to be correct and being followed.
Are all instruments correct?	The instruments had allegedly been calibrated. However, it was observed that a venturi meter was being used in Europe where an orifice meter was used in the United States.
Are laboratory results correct?	All appeared to be correct.
Were there any errors made in original design?	The design was essentially the same as that in the United States with the exception of improvements in the regeneration section.
Were there changes in operating conditions?	No. In fact, the operating conditions of temperature and pressure were identical to those used in the United States.
Is fluid leakage occurring?	This would not explain the problem.
Has there been mechanical wear that would explain problem?	No.
Is the reaction rate as anticipated?	Higher rates of HCl formation could explain part of the problem. On the other hand, the ion exchange bed appeared to be performing as designed except for the frequent regenerations. The frequent regenerations appeared to be associated with the higher-than-anticipated rate of HCl formation.
Are there adverse reactions occurring?	There were no unusual reactions that could explain the problem.
Were there errors made in the construction of the process?	Since the unit had only recently been built, this had to be considered.

The value of using a list similar to that in Table 14-1 is that it helps to eliminate superfluous hypotheses that might be suggested and it allows problem solvers to focus on the likely areas for development of working hypotheses. For example, an examination of this table indicates that areas such as mechanical wear, fluid leakage, or changes in operating conditions are unlikely routes to pursue. In addition, the most obvious conclusion, "the ion exchange resin is not working," is shown by Table 14-1 to be highly unlikely. On the other hand, meter errors, higher-than-expected HCl production rates, and design improvements in the regeneration area appear to be valuable ideas to pursue. As is often the case in industrial problem solving, there may be more than a single item that is causing the problem. For this reason, the problem solver began looking in detail at the design improvements, the variables that might cause the reaction rate to form HCl to be higher than anticipated, and the difference between a venturi meter and orifice meter in this specific application. It was not possible to isolate the potential areas of flow meter error and higher-than-anticipated HCl production rate because the technique for calculating HCl production rate depended on the flow rate of TEG to the ion exchange bed. If this flow meter was in error, the calculated HCl production rate would be in error. Thus both possibilities had to be considered.

Since the operating conditions (temperature and pressure) were identical to those used in the plant in the United States, it would appear that the rate of HCl generation would be the same. However, as indicated earlier, the possibility of higher rates of HCl generation could not be eliminated. Since, essentially, all of the HCl was produced in the first-stage flash drum, the problem solver began an analysis of that operation. Referring back to Chapter 9, the kinetic relationship can be expressed in terms of reaction rate as shown below:

$$R* = C \times DF \tag{14-2}$$

where

$R*$ = rate of change with time per unit volume of the compound under study, mols of HCl/ft^3-min that are formed

C = constant referred to as the "lumped parameter constant"

DF = driving force or incentive for reaction to occur, mols of chlorine derivative gas absorbed/ ft^3 of TEG. Since as indicated earlier essentially all of the HCl generation occurred in the flash drum that was the only area to be considered.

In this case, the constant is the reaction rate for the formation of HCl from the chlorine derivative gas. This will be a function of temperature only. As indicated, the driving force will be the concentration of the gas in the TEG in the flash drum. This will depend on temperature and pressure only. Thus the driving force and lumped parameter constant should be the same in the plants in Europe and the United States. This means that R, the rate of HCl formation in mols per minute per unit of volume, should be the same. Thus if the absolute value of HCl formation in mols/hour is higher, it can only be due to an increase in drum volume or residence time at a constant flow rate. This possibility by itself does not provide enough data to formulate a hypothesis. In order to determine if there is a valid working hypothesis related to increased drum volume, the following items were reviewed:

- The dimensions of the first-stage flash drum were reviewed and it was concluded that the dimensions were identical in the plants in Europe and the United States.
- While the measured liquid levels in the drums were both held at 20%, the ranges of the liquid level instruments were different. This was discovered only during a detailed review of the instrument specification sheets. The instrumentation philosophy in the United States was to only cover the planned range of operations with the liquid level instrument. This would provide a higher degree of accuracy. The range of the level instrument was 50 in. In Europe, the philosophy was to cover the entire height of the drum, 75 in in this case. The Europeans considered this to be a significant design improvement.

Thus the absolute liquid level in Europe was 15 inches versus 10 inches in the United States. Since the drums were the same diameter and the flow rates were identical, this difference would provide 50% more reaction volume in Europe than the United States. This would increase the HCl production by 50%.

In order to determine whether this hypothesis would explain all of the apparent increase in HCl production observed in Europe, an HCl material balance was developed for Europe in a similar fashion to that done for the United States. The premise of these balances was that the HCl removed was equal to the HCl produced. These balances are shown in Table 14.2:

Table 14-2   Chloride balances

Variable	U.S. (base case)	Europe
Flow rate to ion exchange, lb/hr	500	500
Chloride concentration into ion exchange, ppm	500	1500
Chloride concentration out of ion exchange, ppm	0	0
Chloride production by material balance, lb/hr	0.25	0.75
Calculated chloride production, lb/hr[a]	0.25	0.38

[a]The calculated chloride production rate was set at the material balance level for the base case (that in the United States) and was increased by 50% to allow for the increased residence time in Europe.

Thus it appeared that a working hypothesis that the increased residence time was responsible for an increased amount of chloride produced would be a valid working hypothesis. However, the calculated chloride production rate was only about half of the value obtained by material balance (0.38 lb/hr vs. 0.75 lb/hr). The failure to get a good material balance calculation check indicated that there might be other problems. Thus rather than proposing a plant test of lowering the level in the first-stage flash drum, the problem solver began considering other possible problems.

A review of the question guidelines and answers in Table 14-1 indicated that there was likely some mistake in construction or the revised design. Several possible additional hypotheses were developed. Most of them tended to point to the flow measurement of the stream going to the ion exchange bed. If the flow rate was significantly less than that indicated, chlorides would still be removed, but the concentration would build up to a higher level than was anticipated. For example, if the flow was actually 250 lb/hr, the amount of chloride removed in the ion exchange unit would be equivalent to that calculated based on the increased residence in the first-stage flash drum. The problem solver began developing a hypothesis associated with the flow meter by selecting the simplest explanation possible. A review of the venturi meter calculations used for the flow to the ion exchange bed indicated that it was selected to minimize pressure drop. The venturi meter discharge coefficient was assumed to be one which would have been true for a low-viscosity fluid. However, the TEG has a viscosity much higher than that of a typical hydrocarbon. While it appeared to the problem solver that a standard orifice meter would have provided a much more accurate installation, he refrained from indicating this to the Europeans.

The problem solver now developed the following hypothesis that actually contained two possible theoretically correct working hypotheses.

It is believed that the increased absolute level in the flash drum (15 in vs. 10 in) is causing the chloride production to be 50% higher in Europe than in the United States. The level hypothesis does not explain that the concentration of chlorides in the TEG is three times that anticipated based on results in the United States. Thus another problem must be present. It is believed that the measurement of flow to the ion exchange bed is in error.

It should be noted that this problem illustrates that in industrial problem solving, there is often more than one valid hypothesis. Using basic chemical engineering principles can often confirm whether one hypothesis can explain the entire problem.

Step 4: Provide a mechanism to test the hypothesis.

Two separate plant tests were developed to test both of these hypotheses. The tests were conducted concurrently since, as hypothesized, there was no interaction between the two tests. In the first test, the level was reduced in the first-stage flash drum so that the residence time was reduced to 5 min. In the second test, the flow to the ion exchange bed was diverted into a 5-gallon bucket and the actual flow rate was measured. The five gallon bucket required about 11 min to fill, as opposed to the 6 min that it would have required if the flow rate was really 500 lb/hr. In order to confirm that the problem was truly solved, the plant test on the flash drum was continued. That is, the operating directive for the level was set so that the residence time continued at 5 min. In addition, the flow rate to the ion exchange bed was increased to a measured value of 1000 lb/hr. It was anticipated that a measured flow of 1000 lb/hr would give an actual flow of 500 lb/hr. The flow was again measured using the 5-gallon bucket to confirm that a flow rate of about 500 lb/hr was achieved. After a few days at these conditions, the chloride concentration decreased to 500 ppm, the concentration experienced in the United States.

Step 5: Recommend remedial action to eliminate the problem without creating another problem.

The remedial action was relatively simple and consisted in only slight modifi-cations to the changes made to conduct the plant test. These changes were designed to ensure that, at a future time, comparisons made between the two plants did not cause changes in operations which would recreate the startup problems. The changes were as follows:

- The level instrument on the first-stage flash drum was re-ranged from 75 in to 50 in. This would allow both flash drums in the United States and Europe to operate at the same apparent level (20%). This would also give the same absolute level of 10 in. If the level instrument had not been re-ranged, it would have been necessary to maintain the level in Europe at about 13%. There was a concern that this discrepancy between the conditions of the plants in the United States and Europe might lead the European affiliate to raise the flash drum level at some point to be con-sistent with the level in the United States.
- The discharge coefficient for the European venturi flow meter was changed to a value that was based on the actual viscosity, rather than an assumed value of unity.

A detailed potential problem analysis did not reveal any significant new problems if these changes were made.

Lessons Learned If the problem-solving concepts discussed earlier had been applied, the European technical staff might have elected to predict the chloride buildup rate based on data from the United States. If they had been using this potential problem analysis concept, they would have been able to spot the problem much sooner than a month after startup. As can be seen from

Figure 14-2, it was readily apparent after 2 to 3 days that the chloride level was increasing much faster than would have, based on experience in the United States. The advantage of comparing the actual to projected chloride buildup is that problem-solving activity could have started 3 weeks earlier than it actually had.

This problem illustrates the validity of calculations. There will always be a tendency to treat the first discovery as the root cause of the problem. Many industrial problems have more than a single root cause. For example, the discovery that the flash drum in Europe had more residence time than the comparable drum in the United States might have been considered to be the single root cause of the problem. If the problem solver had not concluded, based on calculations, that there must be another cause, the flash drum changes would have been made, but the problem of high chlorides would have continued. In the case described, it was especially desirable to ensure that the problem solutions were complete since the problem solver had only a limited amount of time in Europe.

In our advanced age of electronic equipment, we often forget the more basic measurement techniques. The use of a 5-gallon bucket to measure the low flow rate of a low toxicity and non volatile material is probably one of the best techniques available.

Any design change, regardless of how small (use of venturi meter and change in range of a level instrument), should receive a careful review, including a potential problem analysis.

EXAMPLE PROBLEM 14-2

An Embarrassing Moment

A high vacuum system was designed as part of a new process. The vacuum system was required to achieve an absolute pressure of 15 mm of Hg. In order to do this, a three-stage steam ejector was selected. A schematic of the vacuum system is shown in Figure 14-3.

The construction was relatively straightforward, except for obtaining a steam supply for the ejector. In order to furnish steam and avoid a shutdown of the 200 psig steam supply line, it was necessary to "hot tap" the steam line. Hot tapping is a procedure in which a valve with a flange is welded to the line. When the valve is fully secured and the welding is inspected, a cutting instrument is connected to the open end of the flanged valve. The valve is opened and the cutting instrument is lowered into the valve opening until it touches the pipe. A small pilot drill is first used to cut a small (1/4 to 3/8 in) hole. Following that, a full-size hole is cut. The tool used to cut the full-size hole is then used to pull the piece of the pipe that has been cut out back through the valve. The valve is closed as the tool is removed. The new steam piping to the

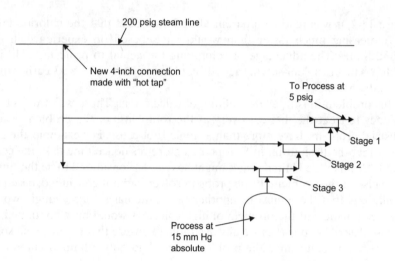

Figure 14-3 Schematic of three-stage steam ejector.

process is then connected to the valve. The valve is then opened at the appropriate point in the startup procedure. The hot tap crew will often save the section
of pipe cut from the main line as proof that the hot tap really has been made.

The startup procedure for the three-stage ejector system called for starting
the stages in the order shown on the schematic drawing. This allowed the
lowest steam demand stages to be started first. The steam jet manufacturer's
guaranteed steam consumptions were as follows:

Stage 1	300 lb/hr
Stage 2	500 lb/hr
Stage 3	1700 lb/hr

The startup procedure seemed to go well, as the first- and second-stage jets
were placed into service. The steam pressure, as measured by a gage in the
new 4-in steam line, remained relatively constant at about 200 psig, and the
pressure in the vacuum drum decreased to the level anticipated with only two
stages in service. However, when steam flow was started to the third stage, the
measured steam pressure on the 4-in line decreased rapidly to 105 psig, and
the pressure on the vacuum drum increased as the first- and second-stage jets
were no longer able to perform at the reduced steam pressure. The initial
reaction of the problem solver was that there was obviously something wrong
with the third-stage jet. After all, the system performed perfectly when only
two stages were used. He called the sales representative and strongly suggested that maybe an orifice had been left out of the third-stage jet. The
problem solver believed that this would cause a huge increase in steam flow
and result in a large pressure drop in the 4-in line. As time passed, the problem
solver began a more methodical approach to analyzing the problem. He used
the five-step approach, as follows.

Step 1: Verify that the problem actually occurred.

There was little doubt that the problem occurred. The drop in steam pressure and the loss of vacuum was also accompanied by loud noises inside the drum as material flowed backwards through the ejectors. However, to satisfy the need to verify the problem and get a maximum amount of data, the startup was repeated. The same results were observed. However, the increased attention to the steam pressure gage on the 4-in line indicated that the steam pressure actually dropped slightly when the first- and second-stage ejectors were placed into service.

Step 2: Write out an accurate statement of what problem you are trying to solve.

The problem solver developed a problem statement as follows:

> During startup of the steam jet system, the steam pressure on the 4-in steam supply line decreased rapidly to 105 psig when the third-stage ejector was placed into service. The operation of the three-stage steam jet is impossible at pressure conditions this low. Very small pressure drops were also observed as the first- and second-stage ejectors were placed into service. The same results occurred during both instances when the ejector system was being placed into service. The pressure on the 200 psig steam pressure header was normal during both trials. There are no steam meters available to measure the actual steam flow to the ejector system. Determine why the steam pressure on the 4-in steam supply line decreased rapidly to 105 psig when the third-stage ejector was placed into service. Recommendations for modifications to allow operating the steam ejector system are also to be provided.

The actual measured pressures from the second trial are shown in Table 14-3.

Step 3: Develop a theoretically sound working hypothesis that explains the problem.

The question list given in Chapter 6 was used as a guide to develop potential working hypotheses. A summary of this analysis is shown in Table 14-4.

Table 14-3   Steam pressure measurements

Ejectors in Service	Steam Pressure, psig
0	200
1	198.5
2	190.3
3	105

Table 14-4   Questions/comments for Problem 14-2

Question	Comment
Are all operating directives and procedures being followed?	The vendor provided startup procedure was being followed exactly.
Are all instruments correct?	The pressure gage that was initially used was replaced with a new gage before the second test.
Are laboratory results correct?	Not applicable in this case.
Were there any errors made in original design?	The 4-in steam line could be too small.
Were there changes in operating conditions?	No. The process was being operated exactly as specified on the duty specification for the steam ejector.
Is fluid leakage occurring?	Not applicable.
Has there been mechanical wear that would explain problem?	Not applicable.
Is the reaction rate as anticipated?	Not applicable.
Are there adverse reactions occurring?	Not applicable.
Were there errors made in the construction of the process?	Since the unit had only recently been built, this had to be considered.

The approach of using the guidelines provided by Chapter 6 may seem trivial for this example, but they helped to isolate the development of a hypothesis to two areas. The areas that the problem solver decided to investigate further were the sizing of the 4-in steam supply line and some sort of construction error. The hypothesis of a construction error would also include the possibility of an error in the manufacturing of the steam ejector. Three potential working hypotheses were proposed, as described below:

1. The process designer had made a mistake in the sizing of the 4-in steam supply line.
2. There was an error made in the construction of the third-stage steam jet.
3. The hot tap crew had made an error and did not completely cut and remove the 4-in piece of the line.

Initially, the hot tap crew was contacted to ascertain that that they did pull a piece of pipe from their cut to determine that the hot tap had indeed been completed. Unfortunately, this was many weeks after the cut had been made and the problem solver was told that all cuts made during that time frame had been discarded.

In order to narrow down the number of working hypotheses to a minimum, the following actions were taken and the indicated results obtained:

- The steam supply line was resized and it was concluded that the 4-in line was more than adequate. The pressure drop should be less than 2 psi even if the steam rates were double the rates guaranteed by the steam jet manufacturer.
- The third-stage jet was removed and inspected and its dimensions were compared to the factory issued drawings. The dimensions of the ejector were as specified in the drawings.

The only remaining hypothesis was that the hot tap had not been completely cut through and that all of the steam was flowing through the pilot drill hole or another restriction in the piping.

Step 4: Provide a mechanism to test the hypothesis.

While the hypothesis could have been tested by insisting that the hot tap crew return and redo their hot tap, the problem solver decided to test the hypothesis using calculations. He proceeded to consider the data shown in Table 14-5.

If the hot tap had not been cut completely, the pressure drop across the pilot drill hole should be proportional to the steam flow rate squared. This is a standard concept of chemical engineering as discussed in Chapter 5. Thus he decided to plot the pressure drop versus the flow rate squared. The resulting plot is shown in Figure 14-4.

The resulting plot had a slope of 0.0000152 and an intercept of zero. Thus it could be specified by the following relationship:

$$\Delta P = 0.0000152 \times F^2 \qquad (14\text{-}3)$$

where

ΔP = pressure drop across the restriction, psi

F = steam rate, lb/hr

Table 14-5 Steam flow and pressure drops

Stages in Service	Steam Flow, lb/hr[a]	Pressure Drop, psi
1	300	2
2	800	10
3	2500	95

[a]The steam flows were taken from the steam jet manufacturer's specification sheet.

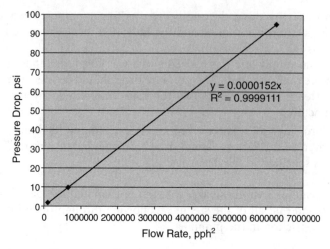

Figure 14-4 Pressure drop vs. flow rate squared.

The intercept of zero was as expected. That is, at no flow there should be no pressure drop. The slope of 0.0000152 was used to approximate the size of the opening that steam was flowing through. To do this, the problem solver used the following relationship for fluid flowing through a restriction. This equation was originally given as equation (5-26) in Chapter 5.

$$\Delta P = 0.5 \times S \times U^2 / 148.2 \qquad (14\text{-}4)$$

where

S = density of the flowing fluid relative to water

U = velocity of the flowing fluid through the orifice, fps; the constants represent conversion factors and the orifice discharge coefficient

Since there was now an experimental relationship between the flow rate and the pressure drop, as well as a similar theoretical relationship, these two equations could be used to estimate the diameter of the restriction. To make this estimate, the right hand side (RHS) of equations (14-3) and (14-4) were set equal to each other since they were both equal to the pressure drop in the restriction.

$$0.0000152 \times R^2 = 0.5 \times S \times U^2 / 148.2 \qquad (14\text{-}5)$$

knowing

$$U = R / (3600 \times A \times \rho) \qquad (14\text{-}6)$$

$$S = \rho/62.4 \qquad (14\text{-}7)$$

where

A = area of restriction, ft^2

ρ = fluid density, lb/ft^3

Solving these two relationships for the area of the restriction and then calculating the resulting diameter gave a value of 3/8 in. This was likely the diameter of the pilot drill used for the hot tap of the steam line.

It appeared likely that the hot tap had not been cut all the way through and that steam was only flowing through the opening that was cut for the pilot drill. When the hot tap crew was contacted, they agreed to return to recut the hot tap only after considerable discussion. When they did recut the hole, they were surprised to find that the piece of material that they removed was exactly as calculated by the problem solver. That is, it was a 4-in piece of metal with a 3/8-in hole in it.

Step 5: Recommend remedial action to eliminate the problem without creating another problem.

No additional actions appeared to be required after the hot tap was recut. The steam ejector system was started up successfully and the steam pressure on the 4-in supply header remained constant at 200 psig throughout the startup and operation.

Lessons Learned   This problem illustrates how jumping to conclusions can often lead to embarrassing moments. Rather than immediately confronting the sales representative, the problem solver should have made a careful study of the available data. A careful study of the data would have revealed that the steam supply pressure did not stay constant even when the smaller first- and second-stage jets were placed into service. The problem also illustrates the value of doing calculations to attempt to pinpoint the problem source. After the designer had rechecked the sizing calculations for the 4-in steam line, the calculations described in Step 4 should have been done. If these additional calculations had been done, there would have been no reason to open and check the dimensions in the third-stage jet. The removal of this third-stage jet was a major effort since it was located three levels up in the structure and had large pipes connected to it.

The calculations described in Step 4 would have been sufficient to point out that the hot tap was likely done incorrectly, and thus would have eliminated the need to remove and inspect the third-stage jet. These calculations also provided a strong argument for redoing the hot tap, as opposed to providing simply a suspicion that it was not done correctly.

EXAMPLE PROBLEM 14-3

Prime Mover Problems Are Not Always What They Appear to Be

An ethylene refrigeration system was expanded by increasing the capacity of a blower. Prior to the expansion, this blower was used to boost the pressure on the system from 10 in of vacuum to 8 psig. The system was expanded by increasing the blower discharge pressure (also the compressor suction pressure) to 10 psig. The increase in pressure to 10 psig was to provide an increase in capacity of about 10%. No other changes were required in the ethylene compression or condensation system. The increased capacity of the blower was to be obtained by replacing the existing impeller with a larger impeller. This increase in impeller size would allow an increased flow rate and an increase in discharge pressure to 10 psig. A short shutdown was required in order to install the new impeller. A schematic drawing of the process is shown in Figure 14-5.

Ethylene liquid flows from the compression and condensation block to provide refrigeration for a low-temperature process operating at approximately –150°F. The ethylene liquid is vaporized at 10 in of mercury vacuum and flows as vapor to the ethylene blower. Prior to the blower, a series of economizers (heat exchangers) raise the temperature of the ethylene from about –156°F to –40°F. The blower boosts the pressure from 10 in vacuum to 10 psig. The ethylene gas at 10 psig flows to the reciprocating compression system where it is compressed to approximately 350 psig and condensed in heat exchangers that are cooled by vaporizing propane. The refrigeration load is not constant. The rate of vapors flowing to the ethylene blower and

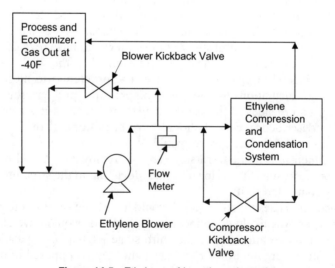

Figure 14-5 Ethylene refrigeration schematic.

compressors varies significantly. To maintain the suction pressure of the blower and compressor, constant "kickback valves" are provided. As the refrigeration load decreases, the kickback valves open, keeping the suction pressures constant.

The operation of the expanded blower was disappointing. While it appeared that additional ethylene flow had been obtained, the pressure was well below the required discharge pressure of 10 psig. Since ethylene flowed from the blower discharge to the compressor suction, this lower discharge pressure also resulted in a lower compressor suction pressure. The lower suction pressure caused both a reduction in compressor capacity and an increase in the compression ratio. Operations personnel requested problem-solving help because, in their words, "This new impeller is not as good as the one that we took out! We never had this kind of trouble before and that kickback valve was always open at least 10%." The problem solver used the five-step approach to assess the situation.

Step 1: Verify that the problem actually occurred.

Verification that there was a problem was relatively easy. When attempts were made to increase the unit production and, hence, the refrigeration load, above that possible before the new impeller was installed, the blower kickback valve would close all the way and the blower suction pressure would increase above the operating value of 10 in of vacuum. Since operating at 10 in of vacuum was necessary in order to maintain the process temperatures, the blower discharge pressure was reduced, causing the reciprocating compressors to have less than the desired capacity.

Step 2: Write out an accurate statement of what problem you are trying to solve.

Prior to attempting to write out an accurate problem statement, the problem solver decided to look at the blower manufacturer's supplied compressor curve for the new impeller. As part of the preparation for this assessment of comparing the theoretical blower curve to actual performance, he had all the key meters checked, so he knew that the blower suction pressure, discharge pressure, and flow rate variables were as accurate as possible. In order to assess the blower performance, it was necessary to maintain the kickback valve in the closed position during the test. This was because the flow meter was located outside the kickback valve line, as shown in Figure 14-5. The problem statement that he developed was as follows:

> The performance of the ethylene blower seems to be worse than anticipated with the new impeller. Rather than obtaining a 10–12% improvement in the plant capacity, operations since the startup of the revised facilities have resulted in a capacity only slightly above the previous capacity. While no test data exists for

performance with the old impeller, operations personnel believe that the performance was adequate when the old impeller was being utilized. They also indicate that the kickback valve was normally open at least 10%. Currently, the valve is closed whenever the system is fully loaded. The ethylene compressors are operating as predicted. The problem is not related to instrumentation, since all the meters have been checked. Determine the following:

- "Is the blower operating as specified by the manufacturer's supplied blower curve?
- "If it is not, determine why.
- "Recommend changes to correct the problem or operating conditions that will allow operation at full capacity."

As indicated in Chapter 3, time is always an important component of a problem statement. In this problem, while no data was available from past operations, the problem solver still noted that performance seemed adequate with the old impeller. In addition, he indicated that the problem seemed to have been present since the startup of the expanded facilities. This helps to focus on the time period after the facilities were expanded. In parallel with developing working hypotheses, the problem solver decided to run a series of plant tests to assess the actual blower performance. A summary of these plant tests is shown in Table 14-6. In addition, the results are shown graphically in Figure 14-6.

Also note in the problem specification that the problem solver takes into account the operator's observations that, prior to the expansion, the perfor-

Table 14-6   Blower capacity tests

Variable	Design	Test 1	Test 2
Molecular weight	28	28	28
Specific heat ratio	1.25	1.25	1.25
Gas compressibility	1	1	1
Polytropic efficiency, %	70	TBD	TBD
Suction pressure, Hg	10	7.5	8.7
Discharge pressure, psig	10	10	10
Gas density at suction, lb/ft^3	0.0609	0.0670	0.0633
Flow rate, lb/hr	35000	42590	36560
ACFM	9580	10590	9630
Temperatures			
Out economizer, °F	−40	−40	−40
Blower suction, °F	−40	−30.5	−29.3
Blower discharge, °F	87	81	91
Blower speed, RPM	10000	10000	10000
Calculated polytropic head, ft	24560	21525	23250
Projected polytropic head, ft		23200	24500
Performance deficiency, %		7.2	5.1

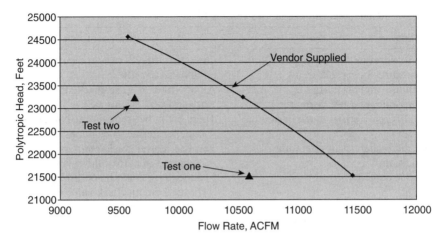

Figure 14-6 Blower capacity curve.

mance was adequate and that the kickback valve was open at least 10%, essentially all of the time. However, the problem statement did not include their conclusion that the blower was performing better with the old impeller.

The polytropic head is calculated using the following equation, given in Chapter 7:

$$H = 1545 \times T_S \times Z \times (R^\sigma - 1)/M\sigma \qquad (7\text{-}6)$$

where

H = polytropic or adiabatic head, ft

T_S = suction temperature, °R

Z = average (suction and discharge) compressibility

R = compression ratio

M = gas molecular weight

σ = polytropic or adiabatic compression exponent

The projected polytropic head is taken from the blower manufacturer's supplied head curve shown in Figure 14-6. The performance deficiency is simply the deviation from the projected head curve, expressed as a percentage.

Figure 14-6 clearly indicates that the blower does not appear to be performing as predicted by the performance curve. However, this data does not by itself provide a working hypothesis. For example, if one simply presents a working hypothesis that says, "The blower is not performing as predicted by the blower manufacturer's supplied capacity curve," does this mean that the blower should be shutdown for maintenance, or is there another problem that

Table 14-7 Questions/comments for Problem 14-3

Question	Comment
Are all operating directives and procedures being followed?	Operating directives were being followed exactly.
Are all instruments correct?	All instruments were checked.
Are laboratory results correct?	Not applicable in this case.
Were there any errors made in original design?	There could be errors in the compressor impeller design calculations.
Were there changes in operating conditions?	Yes. The blower discharge pressure was increased as part of the revised design.
Is fluid leakage occurring?	Fluid leakage could be occurring through the kickback valve or through blower internals.
Has there been mechanical wear that would explain problem?	Blower wear rings are a potential problem.
Is the reaction rate as anticipated?	Not applicable.
Are there adverse reactions occurring?	Not applicable.
Were there errors made in the construction of the process?	Since the unit had only recently been expanded, this had to be considered.

is causing the blower to appear to be operating differently than predicted by the performance curve?

Step 3: Develop a theoretically sound working hypothesis that explains the problem.

The question list given in Chapter 6 was used as a guide to develop potential working hypotheses. A summary of this analysis is shown in Table 14-7. Using these questions from Chapter 6, several hypotheses were developed, as follows:

- There could have been errors in the design calculations for the new compressor impeller. These could consist of either errors in the data supplied to the manufacturer or errors made by the manufacturer.
- An impeller of the wrong size could have been installed.
- The increase in the blower discharge pressure could result in more leakage through wear rings.
- The increase in the blower discharge pressure could cause more leakage through the blower kickback valve.
- The poor insulation on the line between the economizer and the blower suction could cause the ethylene to warm up and thus cause a loss in capacity. As shown in Table 14-6, the gas temperature is increasing from $-40°F$ to about $-30°F$.

All of these are possible hypotheses. The problem solver thought that with additional data and/or calculations that he could eliminate some of them. So he reviewed the original physical properties and design bases and confirmed that they were correct. He then reviewed the purchase order and blower manufacturer's specification for the new impeller and compared them to the bases for the upgraded blower. He found that these were consistent. Of course this does not eliminate the possibility that the wrong impeller was shipped from the supplier. He also compared the old blower curve to the new blower curve and found that they were consistent. That is, when extrapolating from the old blower curve to the new blower curve using the appropriate diameter scaling factors, the extrapolated blower curve was essentially the same as the one supplied by the blower manufacturer. Based on this work, he believed that he had done all that he could do except recommend a blower shut down to eliminate hypothesis 1. Before recommending that the blower be shut down to inspect the impeller to confirm that it was the correct diameter, he decided to consider the other hypotheses.

If there was internal leakage due to excessive clearance inside the blower, the internal gas recirculation would cause a decrease in the polytropic efficiency. This could be determined by the blower suction and discharge temperatures. The following equations from Chapter 7 were used to estimate the efficiency for the two tests. The results are shown in Table 14-8.

$$\sigma = (k-1) \times 100 / (k \times E) \qquad (7\text{-}7)$$

where

E = either adiabatic or polytropic compression efficiency, percent
k = ratio of specific heats, C_p/C_v

$$T_D = T_S \times R^\sigma \qquad (7\text{-}8)$$

where

T_D = absolute discharge temperature
T_S = absolute suction temperature

Table 14-8   Calculated efficiencies for test runs

	Test one	Test two
Suction temperature, °R	429.5	430.8
Discharge temperature, °R	541.0	551.1
Compression ratio	2.24	2.37
Compression exponent	0.286	0.285
Polytropic efficiency, %	70	70.2

Based on the test runs, there does not appear to be any indication of internal leakage, since the calculated efficiencies from the suction and discharge temperatures appear to be essentially the same as the design. Thus hypothesis 3 was eliminated.

As indicated earlier, one hypothesis was that the poor insulation was allowing a 10°F increase in temperature between the economizer and the blower suction. This increased suction temperature would cause an increase in polytropic head. To determine if this was a theoretically sound working hypothesis, the problem solver calculated the blower head, assuming that the gas temperature stayed at −40°F. He obtained the results shown in Table 14-9.

If the gas temperature stayed at −40°F, the required head would have been reduced slightly. The reduced head would decrease the horsepower requirements. However, the system does not appear to be limited by power requirements. As indicated in Table 14-6, the steam turbine driving the compressor remained at the design speed of 10,000 RPM throughout the tests. Thus it appeared that the probability that this hypothesis was correct was very low.

The elimination of these hypotheses left only the alternative hypothesis that there was excessive leakage across the kickback valve. If leakage was occurring through this valve when it was in the closed position, that would explain both the increase in temperature between the economizer and the blower suction, as well as the poor performance of the blower relative to the manufacturer's supplied curve. Rather than immediately recommending a shutdown to inspect the valve, the problem solver reviewed the specifications for the 14 in kickback valve. When he reviewed the specifications, he found that the valve was not specified as a tight shutoff valve. In addition, a review of the drawings indicated that the butterfly valve had a peripheral clearance of 0.05 in. That is, there was a clearance of 0.05 in between the flap of the butterfly valve and the wall of the valve. He then estimated the leakage that could occur across the valve when it was completely closed. The flow rate through this small opening will be at sonic velocity (sonic velocity was described in Chapter 5). The calculations required to estimate the leakage through the valve are as follows:

$$A = \pi \times (D_1^2 - D_2^2)/4 = \pi \times (14^2 - 13.9^2) = 2.19 \text{ in}^2$$
$$= 0.0152 \text{ ft}^2 \tag{14-8}$$

$$P = 0.55 \times (14.7 + 10) = 13.58 \text{psia} \tag{14-9}$$

Table 14-9   Calculation results

	Test one	Test two
Gas rate, ACFM	10590	9630
Polytropic head at suction temperature, ft	21525	23250
Calculated polytropic head at −40°F	21050	22700

$$VS = (P \times g \times k/\rho))^{0.5} (13.58 \times 144 \times 32.2 \times 1.25/0.065)^{0.5}$$
$$= 1100 \text{ ft/sec} \qquad (14\text{-}10)$$

$$F = \rho \times VS \times A = 0.065 \times 0.0152 \times 1100 \times 3600$$
$$= 3910 \text{ lb/hr} \qquad (14\text{-}11)$$

$$ES = F/(60 \times \rho) = 3910/(60 \times 0.065) = 1000 \text{ ft}^3/\text{min} \qquad (14\text{-}12)$$

where

A = peripheral area with a clearance of 0.05 in
D_1 = approximate diameter of valve, in
D_2 = approximate diameter of the butterfly wafer, in
P = pressure at restriction, psia
VS = sonic flow velocity, fps
g = gravity factor, fps^2
k = specific heat ratio
ρ = gas density, lb/ft^3
F = flow rate through peripheral area, lb/hr
ES = approximate volumetric flow, ft^3 /min

A brief review of the sonic flow conditions modeled by equations (14-9) to (14-11) may be appropriate. Essentially all chemical engineering text books discuss this phenomenon in more detail than is possible in this book. Chapter 5 includes a brief discussion of this phenomenon. The velocity across the peripheral opening will be at sonic flow velocity. This is because the pressure after the valve is only about 40% of the pressure before the valve. For a gas with a specific heat ratio (k) of 1.25, sonic flow properties occur if the pressure after a restriction is less than 55% of the pressure before the restriction. If sonic flow conditions are encountered, the maximum flow rate (sonic velocity) that will occur across any size opening with any amount of pressure drop is that which occurs when the outlet pressure is 55% of the inlet pressure. Thus the actual flow rate across the peripheral opening is evaluated at 55% of the absolute inlet pressure (equation 14-9) and at the sonic velocity and density at these conditions (equations 14-10 and 14-11).

As shown above, the estimated leakage through the butterfly valve could account for a capacity loss of approximately 1000 ft^3/min. Referring to Figure 14-6, this difference in suction flow rate could explain the deficiency in performance of the blower.

Step 5: Recommend remedial action to eliminate the problem without creating another problem.

The problem solver was faced with no other reasonable recommendation to make except to shutdown the system and replace the kickback valve with one that had a tight shutoff rating. The potential problems that had to be considered were:

- Was the replacement valve really a tight shutoff valve? Would it be possible to find a 14-in valve that would fit into the space available and not have a peripheral opening similar to that of the existing valve?
- Would the new valve fit without a need for significant piping modifications? Could it be installed with a minimal amount of effort?
- Was there anything else that should be considered prior to a recommended shutdown? For example, should the efficiency of the steam turbine be determined to ensure that it is performing as designed?

Lessons Learned This example problem indicates the value of doing a thorough problem analysis rather than just jumping to the conclusion that the blower is not performing as it was designed. If a complete analysis had not been done, the blower might have been shut down for an inspection or additional insulation might have been added to the blower suction lines in hope that this would improve the performance. Either of these solutions which, on the surface, seemed to make sense, would have delayed finding the leakage in the kickback valve. If the blower had been shut down for an inspection and/or replacement of wear rings without knowledge that the kickback valve was leaking, another shutdown would have been required to replace the kickback valve. The analysis conducted here illustrates the value of doing calculations to prove or disprove hypotheses.

As indicated in Chapter 3, a component of successful plant problem solving is a daily monitoring system that allows for the early detection of problems. This early detection will provide an earlier initiation of problem-solving activities than would occur if the problem were allowed to continue to develop. If such a system had been in place in this example, a plot of "head curve deviation" as defined in Table 14-6 would have likely provided an early signal that there was a performance deficiency. An even better approach to the evaluation of a critical piece of revised equipment is to conduct a performance test as soon after startup as possible.

The problem with the leaking kickback valve could have been detected even earlier than the startup of the expanded facilities. If a plant test had been run prior to the shutdown to expand the plant or if the blower performance had been monitored on a daily basis, the problem with the valve could have been detected prior to the expansion. This would have eliminated the downtime required to replace the valve after the facilities startup.

A conservative approach of inspecting the blower during the recommended shutdown to replace the kickback valve could have been taken. However, this would have required major mechanical work and extended the time of the shutdown. Since the calculated polytropic efficiency was very close to the design value of 70%, it is highly unlikely that this would have been a value-added exercise.

EXAMPLE PROBLEM 14-4

The Value of a Potential Problem Analysis

While the utilization of a potential problem analysis was not emphasized in the previous problems, it would have been of great value in this example problem.

A new fractionation process was designed to minimize cost by eliminating a reboiler, minimizing instrumentation, and maximizing heat integration. The fractionation tower products were a high-purity overhead and a high-purity bottoms stream.

A simplified sketch of the process is shown in Figure 14-7. In the figure, the feed to the tower is fractioned into a high-purity overhead methanol product and a high-purity xylene bottoms product. The heat integration is such that the heat input to the tower consists of a controlled vapor flow of xylene from a furnace. This vapor is the same material as the high purity bottoms product, so that no reboiler is required. That is, the vapors from the furnace are fed directly to the tower to provide heat input. The tower reflux is controlled to maintain a tower temperature profile. In addition to the controlled vapor flow to the bottom of the tower, the vapor output from the furnace is also used to heat the tower feed in exchangers. The xylene vapors condensed in the exchangers then flow to the accumulator as shown in Figure 14-7. The design of the exchangers was such that at full capacity and at the design heat transfer coefficient of the exchanger, the outlet material from the exchanger would be condensed xylene at the boiling point and pressure of the accumulator. That

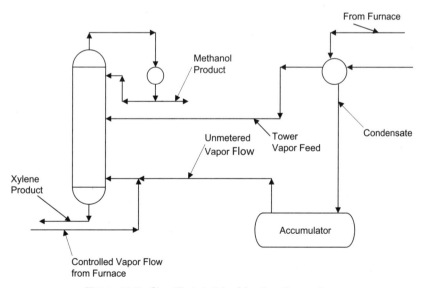

Figure 14-7 Simplified sketch of fractionation system.

is, there would be no significant vapor flow from the accumulator to the tower. The fact that there was no meter on this flow back to the tower did not seem like a significant problem, since there would normally be no flow in the line. However, the designer recognized that there might be times when the conditions were such that the material leaving the exchanger would not be totally condensed and might contain vapor flowing into the accumulator, so he provided a vent line to allow uncondensed vapors from the accumulator to flow to the tower. The designer believed that in cases in which the vapor did vent out of the accumulator in an uncontrolled fashion, the control system would "take care of things" by adding more reflux to the tower.

As often happens, processes that are designed for steady state are rarely operated at steady state. The heat content of the material flowing into the accumulator was constantly changing. At times, there would be two phases (vapor and liquid) entering the drum. In this case, vapor would vent out of the drum into the tower. At other times, the xylene being condensed in the exchanger would be cooled to the point at which the material entering the accumulator would be below the boiling point at accumulator pressure. If this occurred, vapor would flow back out of the tower. The predominant situation was the unsteady state cycling of no vapor being vented back to the tower, transitioning to one where there were two phases in the flow to the accumulator. In this case, where there were two phases present in the flow to the accumulator, vapor would flow uncontrollably to the tower, creating an increase in the heat flow to the tower. The control system would respond after the temperature profile was disturbed and cause more reflux to be added to the tower. During this transient condition, the purity of the overhead product stream would be less than desired, since the increased vapor rate would cause more low-volatility material to be carried overhead until the control system responded and increased the reflux rate. If the reflux rate had to be increased too much to compensate for the vapor venting out of the accumulator, it was possible that the tower would flood. As this situation transitioned to one in which there was no vapor vent from the accumulator, the temperature profile in the tower would again be upset since there was now excessive reflux going to the tower. Again the control system would correct the reflux rate, but only after the bottoms product was off specification.

When the converse situation occurred, the vapor flowed back out of the tower due to the low pressure in the accumulator. In this case, the temperature profile would again be disturbed and the control system would again respond after the disturbance occurred. In this case, it was generally the bottoms stream that would be below specification during the transient.

The problem solver used the five-step procedure discussed earlier to begin solving the problem.

Step 1: Verify that the problem actually occurred.

The initial description of the problem was only that something was causing an upset in the tower and the operations people believed that it was somehow associated with the accumulator. They often tried to compensate for these upsets by trying to adjust the controlled vapor rate. However, this was largely guess work and often made things worse. The problem solver verified that upsets in the tower were being caused by changes (increases or decreases) in the fraction of vapor in the condensate flowing to the accumulator.

Step 2: Write out an accurate statement of what problem you are trying to solve.

The problem solver wrote out the following problem description:

> Fractionation tower upsets are being caused by changes in an unmetered flow going to the bottom of the tower. These changes in the unmetered flow cause an increase or decrease in heat input to the bottom of the tower, the temperature profile in the tower to be upset, and the purity of the distillate and bottom products to be off specification. Determine how to eliminate the fractionation tower upsets caused by changes in the unmetered flow going to the bottom of the tower.

Step 3: Develop a theoretically sound working hypothesis that explains the problem.

In this example, a start of the working hypothesis that explains the problem was included as part of the problem statement. However, it was not obvious how to solve the problem until the hypothesis was more fully developed. The questions given in Chapter 6 were used to fully develop a working hypothesis for obtaining a solution to the problem, as shown in Table 14-10.

Table 14-10 Questions/comments for Problem 14-4

Question	Comment
Are all operating directives and procedures being followed?	All operating directives and procedures were being followed. New ones were considered, but would not solve the problem.
Are all instruments correct?	Yes.
Are laboratory results correct?	Not applicable in this case.
Were there any errors made in original design?	The assumption of steady state was not valid.
Were there changes in operating conditions?	No.
Is fluid leakage occurring?	Not applicable.
Has there been mechanical wear that would explain problem?	No.
Is the reaction rate as anticipated?	Not applicable.
Are there adverse reactions occurring?	Not applicable.
Were there errors made in the construction of the process?	Since the unit had only recently been built, this had to be considered.

The only two reasonable hypotheses were that the assumption of steady state operations was not valid and the possibility that a construction error had been made. No specific hypothesis was developed that would tie construction errors to the symptoms being observed. While it was possible that an error in the tray design, fabrication, or installation might be possible for fractionation upsets at the extreme conditions of high rates of uncontrolled vapors to the tower, it seemed unlikely that these tray errors would not show up at other times. Exploring the construction error hypothesis would likely require elaborate test equipment and/or a tower shutdown. It was decided to first consider the possibility that the original assumption of steady state operation was the primary cause of the problem. It was clear that if the temperature of the condensate returning to the drum was not at the boiling point at the pressure in the drum, there would be an unmetered flow either to or from the tower. Developing the simplest solution for the problem, as pointed out in Chapter 3, is always the best approach.

Thus the problem solver developed the following hypothesis:

> It is believed that the problems associated with the control of the tower are due to the fact that the heat content of the stream leaving the exchangers is not constant. At times, there are large amounts of vapor in this stream which then vent to the tower as an uncontrolled heat input. At other times, the stream leaving the exchangers is subcooled, which causes vapors to flow from the tower to the accumulator. Tower control will be greatly improved if the vent or back flow from the accumulator can be measured.

Step 4: Provide a mechanism to test the hypothesis.

As what was thought to be a permanent solution to the problem, a venturi meter was installed in the vapor line. The venturi meter was selected because it would have a low pressure drop and because it had the inherent capability to measure flow in both directions. If the enthalpy of the flow to the accumulator was such that some flashing occurred in the accumulator, the venturi would measure flow from the accumulator into the tower and the controlled vapor rate from the furnace would be reduced to compensate for this vapor flow from the accumulator. Thus the vapor rate in the tower would remain constant. Conversely, if the enthalpy of the flow into the accumulator was such that the liquid in the accumulator was subcooled, creating back flow from the tower, the controlled vapor rate would be increased to compensate for this back flow. It was believed that the installation of the venturi meter would maintain the vapor flow in the tower constant and thus avoid tower upsets. The control algorithm for the system is described as shown below:

$$F = V - Y + ZF \qquad (14\text{-}13)$$

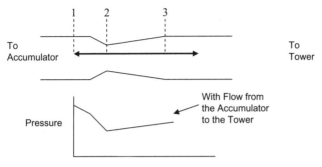

Figure 14-8   Venturi sketch.

where

   V = tower internal vapor rate, which should be held constant
   F = controlled vapor rate from the furnace
   Y = flow rate from the accumulator to the tower
   ZF = flow rate from the tower to the accumulator

   A simplified sketch of the venturi meter design is shown in Figure 14-8. A typical venturi pressure profile is also shown for the case where there is flow from the accumulator to the tower. It was recognized that the accuracy of the flow from the accumulator to the tower (measured by the pressure drop from points 1 to 2) would be more accurate than the backflow from the tower to the accumulator (measured by the pressure drop from points 3 to 2).

   Unfortunately, no potential problem analysis (as suggested in Chapter 3) for this problem solution was done. The fact that there is pressure recovery with any type of meter was not considered. This pressure recovery is represented in the sketch as the pressure increase from point 2 to point 3, when flow is from the accumulator to the tower. Since the pressure at point 3 is greater than the pressure at point 2, the control system would assume that this was backflow from the tower. The control system would then have values for both Y and ZF. Of course, when flow was from the accumulator to the tower, the actual value of ZF was zero. However, the pressure recovery made the control system think that ZF had a nonzero value. A potential problem analysis that included a detailed understanding of the venturi meter would have discovered this problem and allowed an engineering solution well before startup of the revised facilities.

Step 5: Recommend remedial action to eliminate the problem without creating another problem.

After the initial startup difficulties discussed above, a selector switch was installed to allow the control scheme to select the greater value of Y or ZF

and set the other variable to zero. The system performed flawlessly after that minor modification.

Lessons Learned   There are several lessons that can be learned from this problem. While process design involves the assumption of steady state, consideration should always be given to the question of "How does unsteady state impact the design?" If the process designer had considered unsteady state, it is likely that he would have provided flow measuring devices as part of the original design. The problem solver can also use the question of "How does unsteady state impact things?" as a problem-solving tool by questioning the validity of the steady state assumption.

There is great value in both understanding the equipment involved in a problem solution and in performing a potential problem analysis prior to making a recommendation. In the example given here, the fact that pressure recovery would impact the results was blatantly obvious to anyone with a minimal knowledge of flow instruments. However, in the rush to get the facilities designed and installed, it was overlooked. The discipline to conduct a potential problem analysis would have pinpointed this problem before the venturi meter was installed. Potential problem analyses are often not done except if they are required as part of a disciplined procedure.

NOMENCLATURE

A Area. In this chapter it is used to represent the area of a restriction or a peripheral area with a clearance of 0.05 in. The value is in ft^2.

C A constant referred to as the "lumped parameter constant"

D_1 Approximate diameter of valve, in

D_2 Approximate diameter of the butterfly wafer, in

DF Driving force or incentive for reaction to occur, mols of chlorine derivative gas absorbed/ft^3 of TEG

E Either the adiabatic or polytropic compression efficiency, %

ES Approximate volumetric flow, ft^3/min

F Flow rate. In this chapter, it is used to represent the ion exchange bed feed rate, flow rate through a peripheral area, the flow of steam through the restriction, or the controlled vapor rate from the furnace, all in lb/hr.

g Gravity factor, fps^2

H Polytropic or adiabatic head, ft

k Ratio of specific heats, C_p/C_v

M Gas molecular weight

P Pressure at restriction, psia

R Compression ratio

$R*$ Rate of change with time per unit volume of the compound under study, mol of HCl/ft^3-min that are formed

RC Rate of chloride production in the United States, lb/hr

S Density of the flowing fluid, relative to water

T_D Absolute discharge temperature

T_S The suction temperature, °R

U Velocity of the flowing fluid through the orifice, fps

V Tower internal vapor rate which should be held constant

VS Sonic flow velocity, fps

X_F Concentration of chlorides in the flow to the ion exchange bed, weight fraction

Y Flow rate from the accumulator to the tower

Z Average (suction and discharge) compressibility

ZF Flow rate from the tower to the accumulator

ΔP Pressure drop across the restriction, psi

ρ Fluid density, lb/ft³

Σ Polytropic or adiabatic compression exponent

15

A FINAL NOTE

Since problem solving as described in this book is not an intuitive process, there will be a learning period or an induction period before it can be done efficiently. Because of this learning period, the first few problems or calculations that are done using this process will be slow and labor intensive. My personal assessment is that the first time a new calculation technique is utilized, it will often take five to ten times as long to do the calculation as it will after the calculation technique becomes second nature. This is because the problem solver often will not have a feel for what the magnitude of the value that he is calculating should be and he will conservatively check his calculations against some known standard. However, as experience is gained with this technique, speed will increase and some of the labor intensive steps will become almost automatic. Best of all, problems will be solved one time for all time. In addition, as indicated in earlier chapters, this approach is not to be used for every plant problem-solving activity. There will always be questions of optimum technical depth to consider.

There are some guidelines that should be considered prior to beginning to either implement these techniques as an individual or, as a manager, to implement these techniques throughout an organization. Some of these guidelines have been mentioned in earlier chapters. The purpose of this chapter is to present some of these guidelines together in one location.

The problem solver and his management should realize that the first time these techniques are used, more time will be required. The process of using a list of questions such as those in Chapter 6 to develop a working hypothesis will be cumbersome until it is done a few times. There will be calculation

Problem Solving for Process Operators and Specialists, First Edition. Joseph M. Bonem.
© 2011 John Wiley & Sons, Inc. Published 2011 by John Wiley & Sons Inc.

techniques that the problem solver is not familiar with. He may want to check his calculations because he has not yet developed an intuitive feel for the particular calculation. Management can be a great assistance in this area by insisting that they want the problem solver to take the time to do it right and insist on using the techniques described in this book. Our culture demands quick answers in almost all areas. From video games to interactive, computer-aided learning, we have become an iterative society. The high school student who is using the computer to study can often keep trying to get the answer without any thought or calculation until he gets it right. It will take strong management action to insist on replacing this "quick answer" culture with a "do it right" culture.

The data that an operator or mechanic has are invaluable in developing a theoretically sound working hypothesis. They have first-hand experience and have made observations that are not available through any other means such as process control computers, instrumentation, or laboratory results. They may at times have problems putting their observations or experiences into quantitative descriptions.

There will be times when it is necessary to make assumptions. Many times in industrial problem solving, the direction of the change and the order of magnitude of the impact is all that is required. An exact value may not be required. For example, it may be necessary to assume the constant in the relationship between the rate and the driving force. The problem solver should not be discouraged by the need to make an assumption. It is often more important to get the form of the driving force correct than it is to get the constant correct. As described in Example Problem 14-1, knowing the relationship between chloride production and residence time was much more important than knowing the actual rate constant.

Bureaucracy should be avoided at all costs. There will be a tendency for management to micromanage the exact wording of the document described in Table 3-3. This problem specification is meant to be used by the problem solver to provide the best description possible of the problem that he is trying to solve. Wording changes will no doubt end up diluting and confusing his efforts. On the other hand, a review of the final conclusions and proposals to conduct tests to prove the hypothesis is mandatory. These reviews should focus on the technical accuracy of the calculations and the preparation for any proposed plant tests. Safety should be a major component of this review.

The concept of "one riot, one ranger," as described in Section 6.6, should be utilized. While committees are of value, management should make it absolutely clear who has responsibility for developing a problem solution.

There may be value in initially having an individual assigned as the sponsor for utilization of these problem-solving procedures. This individual would be available to consult in the utilization of each of the procedural steps and calculation techniques. He would not necessarily be an expert in the process of interest or in the utilization of equipment calculations. However, he would have sufficient knowledge of the techniques to ensure that all the steps in the process were adequately considered.

APPENDIX

CONVERSION FACTORS

To convert from English units to CGS units, use the following table.

English Units	CGS Units	To Convert to CGS, Multiply by
Feet (f)	Meter (m)	0.3048
Pounds (lb)	Kilograms (kg)	0.454
Pounds/inch^2 (psi)	Bar (bar)	0.069
Gallons (g)	Liters (L)	3.785
BTU	Calories (cal)	252.0
Centipoise (cP)	Poise (P)	0.01
Horsepower (BHP)	Kilowatt (kw)	0.746
Degrees Fahrenheit (°F)	Degrees Centigrade (°C)	(°F − 32) × 0.5556
Degrees Rankin (°R)	Kelvin (K)	0.5556

Problem Solving for Process Operators and Specialists, First Edition. Joseph M. Bonem.
© 2011 John Wiley & Sons, Inc. Published 2011 by John Wiley & Sons Inc.

REFERENCES

Henshaw, Terry L. "Reciprocating Pumps." *Chemical Engineering* (September 21, 1981): 105.

Scheel, Lyman F. *Gas Machinery*. Houston, TX: Gulf Publishing, 1972.

Scheel, Lyman F. "New Ideas on Centrifugal Compressors, Part 1." *Hydrocarbon Processing* 47, no. 9 (September 1968): 253.

Seymour, Raymond B., and Charles E. Carraher. *Seymour/Carraher's Polymer Chemistry: An Introduction*. New York: Marcel Dekker, 1996.

INDEX

Problem Solving for Process Operators and Specialists, First Edition. Joseph M. Bonem.
© 2011 John Wiley & Sons, Inc. Published 2011 by John Wiley & Sons Inc.